Orthacean and strophomenid brachiopo␣ Lower Silurian of the central Oslo Region

B. GUDVEIG BAARLI

Baarli, B.G. 1995 10 30: Orthacean and strophomenid brachiopods from the Lower Silurian of the central Oslo Region. *Fossils and Strata*, No. 39, pp. 1–93. Oslo ISSN 0300-9491. ISBN 82-00-37659-1.

Brachiopods of the Superfamily Orthacea and the Order Strophomenida from the Llandovery of the Oslo–Asker District of Norway are described taxonomically. The fauna comprises 41 species belonging to 29 genera. Among these, 3 genera, 10 species and 1 subspecies are new. They include the genera *Orthokopis*, *Crassitestella* and *Eocymostrophia*, the species *Orthokopis idunnae*, *Eridorthis vidari*, *?Plectorthis* sp., *Skenidioides worsleyi*, *Skenidioides hymiri*, *Dactylogonia dejecta*, *Eostropheodonta delicata*, *Mesopholidostrophia sifae*, *Eocymostrophia balderi*, *Fardenia oblectator* and *Eoplectodonta transversalis* (Wahlenberg, 1818) *jongensis*. Parts of the brachiopod fauna occurring across the Ordovician–Silurian boundary are unique in representing relict Ordovician taxa. They are related to relatively rare offshore faunas preserved in uppermost Ordovician strata; they survived into the Silurian in the tectonically active, deeper-water environments found in the intracratonic basin of the Oslo Region. The rest of the Llandovery fauna shows very close ties with the faunas and faunal structure of the Welsh Basin, suggesting minimal geographic separation between Baltica and Avalonia. Similarities at species level between the Gornyj Altaj and the Norwegian–Welsh fauna also suggest that the two areas were situated in the same climatic zone, contrary to some recent paleogeographic reconstructions. A rare fauna of the Southern Midcontinent of Laurentia lived in similar facies as the Norwegian–Welsh fauna and show strong resemblances on a generic level. Except for some eurytopic species, however, many of the species are different, indicating that Silurian barriers existed between these locations. □*Brachiopods, systematics, Silurian, paleogeography, Norway*.

B. Gudveig Baarli, Department of Geology, Williams College, Williamstown, MA 01267, USA; 9th November, 1993; revised 23rd June, 1994.

Contents

Introduction 3
Stratigraphy and sampling 3
Localities 4
   Asker County 4
   The city of Oslo 6
Measurements and statistics 6
Geographic relationships of the fauna 7
   Comparison between Llandovery faunas 7
   Geographic separation of the Llandovery faunas 7
   The boundary fauna 8
Systematic paleontology 8
   Class Calciata Popov, Bassett, Holmer & Laurie, 1993, Subclass
      Articulata Huxley, 1869, Order Orthida Schuchert &
      Cooper, 1932, Suborder Orthidina Schuchert & Cooper, 1932,
      Superfamily Orthacea Woodward, 1852, Family Orthidae
      Woodward, 1852, Subfamily Orthinae, Woodward, 1852 9
   Genus *Orthokopis* gen. nov. 9
     Orthokopis idunnae sp. nov. 9

Family Dolerorthidae Öpik, 1934, Subfamily Dolerorthinae
   Öpik, 1934 10
Genus *Dolerorthis* Schuchert & Cooper, 1931 10
   Dolerorthis sowerbyiana (Davidson, 1869) 10
   Dolerorthis aff. *sowerbyiana* (Davidson, 1869) 11
   Dolerorthis aff. *psygma* Lamont & Gilbert, 1945 12
Genus *Schizonema* Foerste, 1909 13
   Schizonema subplicatum (Reed, 1917) 13
Subfamily Hesperorthinae Schuchert & Cooper, 1931 14
Genus *Hesperorthis* Schuchert & Cooper, 1931 14
   Hesperorthis hillistensis Rubel, 1962 15
   ?Hesperorthis gwalia (Bancroft, 1949) 16
Subfamily Glyptorthinae Schuchert & Cooper, 1931 17
Genus *Eridorthis* Foerste, 1909 17
   Eridorthis vidari sp. nov. 17
Glyptorthinae sp. 18
Family Plectorthidae Schuchert & Le Vene, 1929, Subfamily
   Plectorthinae Schuchert & Le Vene, 1929 19

Genus *Plectorthis* Hall & Clarke, 1892 19
   ?*Plectorthis* sp. 19
   Plectorthid indet. 19
Subfamily Platystrophiinae Schuchert & Le Vene, 1929 20
Genus *Platystrophia* King, 1850 20
   Platystrophia brachynota (Hall, 1843) 20
Family Skenidiidae Kozłowski, 1929 21
Genus *Skenidioides* Schuchert & Cooper, 1931 21
   Skenidioides worsleyi sp. nov. 21
   Skenidioides scoliodus Temple, 1968 22
   Skenidioides hymiri sp. nov. 23
   Skenidioides sp. 24
Order Strophomenida Öpik, 1934, Suborder Strophomenidina
   Öpik, 1934, Superfamily Plectambonitacea Jones, 1928,
   Family Leptestiidae Öpik, 1933 24
Genus *Leangella* (*Leangella*) Öpik, 1933 24
   Leangella scissa (Davidson, 1871) *triangularis*
     (Holtedahl, 1916) 24
Family Xenambonitidae Cooper, 1956, Subfamily Aegiromeninae
   Havlíček, 1961 26
Genus *Aegiria* (*Aegiria*) Öpik, 1933 26
   Aegiria norvegica Öpik, 1933 26
Family Sowerbyellidae Öpik, 1930, Subfamily Sowerbyellinae Öpik,
   1930 27
Genus *Eoplectodonta* (*Eoplectodonta*) Kozłowski, 1929 27
   Eoplectodonta duplicata (J. de C. Sowerby, 1839) 27
   Eoplectodonta transversalis (Wahlenberg 1818) *jongensis*
     subsp. nov. 29
   ?*Sowerbyella* sp. 31
Superfamily Strophomenoidea Öpik, 1934, Family Strophomenidae
   King, 1846, Subfamily Furcitellinae Williams, 1965 31
Genus *Katastrophomena* Cocks, 1968 31
   Katastrophomena woodlandensis (Reed, 1917) 31
   Katastrophomena penkillensis (Reed, 1917) 32
Genus *Dactylogonia* Ulrich & Cooper, 1942 34
   Dactylogonia dejecta sp. nov. 34
Family Rafinesquinae Schuchert, 1893, Subfamily Leptaeninae Hall &
   Clarke, 1894 36
Genus *Leptaena* Dalman, 1828 36
   Leptaena haverfordensis Bancroft, 1949 36

Leptaena valida Bancroft, 1949 37
Leptaena valentia Cocks, 1968 38
Leptaena purpurea Cocks, 1968 39
Leptaena sp. 39
Crassitestella gen. nov. 39
   Crassitestella reedi (Cocks, 1968) 39
Genus *Cyphomenoidea* Cocks, 1968 40
   Cyphomenoidea wisgoriensis (Lamont & Gilbert, 1945) 40
Family Leptostrophiidae Caster, 1939 42
Genus *Eostropheodonta* Bancroft, 1949 42
   Eostropheodonta multiradiata? Bancroft, 1949 42
   Eostropheodonta delicata sp. nov. 43
Genus *Palaeoleptostrophia* Rong & Cocks, 1994 44
   Palaeoleptostrophia ostrina? (Cocks, 1967) 44
Genus *Mesoleptostrophia* Harper & Boucot, 1978 45
   ?*Mesoleptostrophia* sp. 45
Genus *Eomegastrophia* Cocks, 1967 45
   Eomegastrophia spp.? 45
Family Eopholidostrophiidae Rong & Cocks, 1994 46
Genus *Mesopholidostrophia* Williams, 1950 46
   Mesopholidostrophia sifae sp. nov. 46
Genus *Eopholidostrophia* Harper, Johnson & Boucot, 1967 47
   Eopholidostrophia spp. 47
Family Amphistrophiidae Harper, 1973, Subfamily
   Mesodouvillininae Harper & Boucot, 1978 48
Genus *Eocymostrophia* gen. nov. 48
   Eocymostrophia balderi gen. and sp. nov. 48
Family Strophonellidae Hall, 1879 49
Genus *Strophonella* Hall, 1879 49
Subgenus *Eostrophonella* Williams, 1950 49
   Strophonella (*Eostrophonella*) *davidsoni* (Holtedahl, 1916) 49
Superfamily Davidsoniacea King, 1850, Family Fardeniidae
   Williams, 1965 50
Genus *Saughina* Bancroft, 1949 50
   Saughina pertinax (Reed 1917) *gentilis* (Bancroft 1949) 50
Genus *Fardenia* Lamont, 1935 51
   Fardenia oblectator sp. nov. 51
   Fardiniidae indet. 53
References 53
Plates 56

Introduction

Articulate brachiopods are a conspicuous part of the Llandovery fossil fauna in the central Oslo Region, constituting more than 70% of all macrofossils. This fauna was primarily collected and described for paleoecological studies (Baarli 1987, 1990a). During these studies it became clear that some sections traditionally regarded as entirely Silurian, encompassed the Ordovician–Silurian boundary. An unusual and very rich shelly fauna occurs across this boundary (Baarli & Harper 1986). The dominant group is represented by the orthides. Detailed taxonomic studies of the orthide super-family Enteletacea from the Llandovery of the central Oslo Region, including the Ordovician–Silurian boundary interval, was undertaken by Baarli (1988a). These Enteletacean species are listed in Table 1. Again, the uniqueness of the boundary fauna was further emphasized in that many of the taxa were new and often extended the known stratigraphic range of the genera. The present contribution continues the taxonomic approach and includes descriptions of the orthide superfamily Orthacea and the order Strophomenida from the same sections. This material has enabled identification of 41 different taxa, including 3 new genera and 10 new species, of which 5 are orthides.

Previous work on the Orthacea in the Oslo Region consists only of preliminary taxa lists with figures given by Cocks & Baarli (1982) and Thomsen & Baarli (1982). The strophomenides are better known. In addition to the two lists mentioned above, Holtedahl (1916) described the Paleozoic strophomenides from the entire Oslo Region. Öpik (1933), in his classic treatment of the Plectambonitacea, added to this work. Obviously these references are old and in need of revision but the majority of the strophomenide taxa found in this study were described or figured by Holtedahl (1916).

The faunas occur in mixed limestone and siliciclastic strata deposited in an intracratonic basin, mainly in an offshore setting of moderate water depth (Baarli 1985, 1988b, & 1990b, Möller 1989). Both sea level changes and tectonic disturbances stimulated frequent fluctuations in water depth and lithological changes leading to a very varied and rich brachiopod fauna for the relatively short time period and small area investigated.

Acknowledgement and repository abbreviations. – This work was made possible through a grant from the Nansen Funds of Norway. Professor D. Bruton and the staff at the Paleontologisk Museum in Oslo kindly gave me technical help and let me use their equipment. I am grateful to Dr. Rong Jia-yu for carefully criticizing the manuscript and making many useful suggestions. Dr. D.A.T. Harper made a very thorough review of the manuscript that led to great improvements. Dr. A. Boucot shared discussions on the genus *Eocymostrophia*. Finally I thank my husband, Dr. M.E. Johnson, for help in improving the English language and general encouragement. The entire collection is deposited in the Paleontological Museum of Oslo, Norway. Repository abbreviations are as follows: PMO = Paleontologisk Museum in Oslo; B = Natural History Museum, London; BR = Institute of Geology, Tallinn, Estonia; BU = Birmingham University; GMS = Institute of Geological Science, London; HML = Hunterian Museum, Glasgow; OUM = Oxford University Museum; SM = Sedgwick Museum, Cambridge; USNM = United States National Museum of Natural History, Smithsonian Institution.

Stratigraphy and sampling

The bio- and lithostratigraphical scheme for the Llandovery of the Oslo Region was reviewed by Baarli & Johnson (1988). The lithostratigrahic framework for the central Oslo Region with correlation of the stages is shown in Fig. 1.

Sampling was done in the Asker, Bærum and Malmøya areas of the Oslo–Asker District (Fig. 2).

Most of the fossil material was retrieved by bulk sampling. The 10 kg samples were broken up and treated with 8% diluted hydrochloric acid, which left good internal moulds. The moulds were further enhanced using an engraving 'vibro-tool'. For all species, at least one set of casts of moulds for the internal and external of dorsal and ventral valves was taken, if available, with a latex rubber solution. Very delicate casts were taken with silicon-latex placed in a pressure tank over the night.

This kind of treatment was only possible in the Solvik Formation, where samples were taken at least every 10 m stratigraphically, and in a few shale intervals of the uppermost Rytteråker and basal Vik formations. The mainly calcareous Rytteråker Formation, though quantitatively rich in fossils, is therefore not so well represented. Neither is the

Table 1. This list gives the names of the part of the Llandovery brachiopods fauna from of the central Oslo Region described earlier by Baarli (1988); the Enteletacea Waagen 1884. The names of members and formations are given. Basal Myren indicates that the species are found occuring across the Ordovician–Silurian boundary.

Name	Member	Formation
Dalmanella cf. *pectinoides* Bergström 1968	Padda, Myren	Solvik
Ravozetina cf. *honorata* (Barrande 1879)	basal Myren	Solvik
Isorthis mackenziei Walmsley *in* Boucot *et al.* 1966		Vik
Isorthis prima Walmsley & Boucot 1975	basal Myren–Leangen, Padda	Solvik
Levena sp.	basal Myren	Solvik
Resserella matutina Baarli 1988	Leangen	Solvik
Mendacella bleikerensis Baarli 1988	Spirodden, Leangen	Solvik
Mendacella sp.		Vik
Marklandella markesi Baarli 1988	Leangen	Solvik
Kampella guttula Baarli 1988		Vik
?*Paurorthis inopinatus* Baarli 1988	basal Myren	Solvik
Dicoelosia osloensis Wright 1968	Myren–Leangen, Padda	Solvik
Dicoelosia alticavata (Whittard & Barker 1950)		Vik
Epitomyonia sp.	basal Myren	Solvik
Drabovia sp.	basal Myren	Solvik
?*Diorthelasma semotum* Baarli 1988	Leangen	Solvik
Salopina pumila Baarli 1988		Vik
Chrustenopora askerensis Baarli 1988	basal Myren	Solvik
Jezercia rongi Baarli 1988	basal Myren	Solvik

calcareous and seemingly unfossiliferous Vik Formation with the exception of the shale intervals mentioned above. Spot samples were taken in all formations, and these yielded additional information.

Localities

The sections are listed according to the numbers on Fig. 2 and the NATO reference grid numbers in each of the three areas. These grid numbers (Grid ref.) were taken from map series M 711 sheet 1814 I and 1914 IV from Norges geografiske oppmåling (1971). Localities below are given with grid references from south to north in the different counties. The the method of sampling is also described for each locality.

Asker County

1   *NM 817 328 Skytterveien.* – On the north side of the road, a continuous road section exposing, from the west to the east, the uppermost part of the Myren Member, the entire Spirodden Member and the first 20 m of the Leangen Member, all in the Solvik Formation. Thereafter, approximately 20 m of the profile is covered by soil, followed by a 33 m thick section of the uppermost Leangen Member, which ends close to the base of the overlying Rytteråker Formation. The entire section is

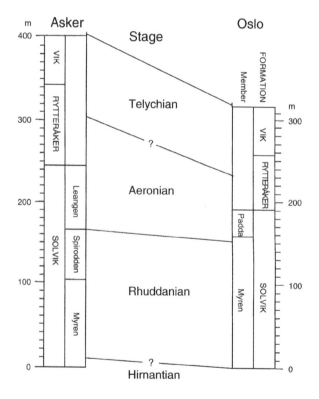

Fig. 1. Stratigraphic sections from the Asker and Oslo areas, showing lithostratigraphic and chronostratigraphic correlations. The section termed Asker is composite with the Solvik Formation measured in Asker county and the Rytteråker and Vik formations measured in Bærum county. There seem to be only small differences in lithology and thicknesses between the Llandovery sections in the two counties.

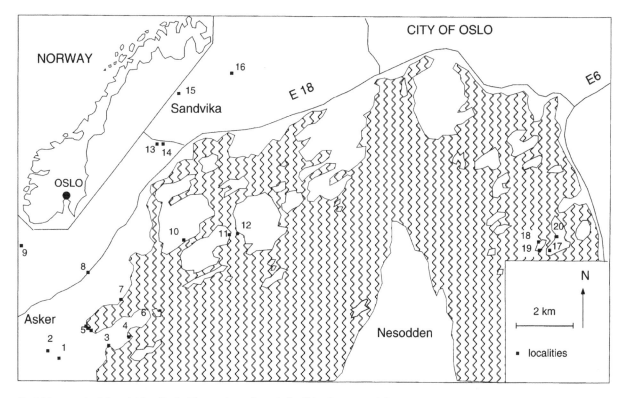

Fig. 2. Map over the Oslo and Asker district. The numbers refer to the localities given on pp. 4–6.

nearly vertical and inverted. Bulk samples were systematically taken every 5 m where the Leangen Member is exposed. The rich fossil occurrences and excellent preservation also encouraged extensive spot sampling.

2   *NM 814 329 Bleikerveien.* – This road section lies near the farm Øvre Bleiker. Approximately 10 m of the uppermost Leangen Member of the Solvik Formation is followed by 30–40 m of the Rytteråker Formation. Only spot sampling near the base of the Rytteråker Formation was undertaken.

3   *NM 835 333 Vettrebukta.* – East of the farm Vettre, immediately northeast of the little bay, is a beach section. The strata trend nearly parallel with the coast and consist of the lime-rich Spirodden Member forming inland cliffs, and the first 5–7 m of the shaly Leangen Member where the section ends in the sea. Fossils were collected only from Leangen Member by extensive spot sampling.

4   *NM 839 338 – 841 339 – 839 336 Spirodden.* – This is the most complete and longest section of the Solvik Formation. A 160 m long beach section of near vertical strata is exposed along the coast from a point on the south side of the peninsula Konglungen. The section exposes 104 m of the Myren Member, the entire Spirodden Member, and 4–5 m of the Leangen Member. Spot sampling was difficult because of the exposure to the seawater of visible

fossils. The bulk samples were taken every 10 m, except near the base of the Myren Member where the distance was 5 m.

5   *NM 825 341 – 825 343 Leangbukta.* – This road section exposes 75–80 m of the Leangen Member, Solvik Formation, with only the base missing. The profile continues into the Rytteråker Formation, which can be found in the hillside above the road. There is, however, some tectonic disturbance near the top of the Leangen Member, making the total thickness of the member slightly uncertain. Ten bulk samples were taken throughout the section in addition to some spot sampling.

6   *NM 849 347 Konglungø.* – The Ordovician–Silurian boundary is exposed here on the beach on the south side of the small island Konglungø. The island is connected by a bridge to the peninsula Konglungen. The oolitic limestones of the Ordovician Langøyene Sandstone Formation are directly overlain by about 12 m of shales of the Myren Member, Solvik Formation. The boundary occurs somewhere within those 12 m. Bulk samples from the base and 3 and 9 m above the base of the member were obtained together with extensive spot sampling.

7   *NM 837 352 Hvalstrand.* – This is a public beach, where rock exposures occur in the northernmost part. The boundary between Langøyene Sandstone Formation

and the Solvik Formation is exposed together with the first few meters of the Myren Member. Only minor spot sampling was undertaken.

8   *NM 828 357 Vakås.* – The exposure is on the west side of the highway E18 from Oslo to Drammen. The Silurian section is approximately 100 m thick and runs through most of the Myren Member, Solvik Formation. Only four bulk samples were collected from the basal layer to a level 17 m above the base of member. No spot sampling was done owing to dust and the danger from traffic. Special permission from the highway authorities is required to work in this section. The bulk samples were very rich in fossils.

9   *NM 805 367 Semsvannet.* – No exact location is given for the specimens found in the auxiliary collection in the Paleontological museum in Oslo. Professor Kiaer (1908), who collected most of the specimens, mentioned, however, an especially good profile between Sem and Tangen. This section is part of a syncline where lowermost Llandovery rocks of the basal Myren Member, Solvik Formation, are exposed between uppermost Ordovician strata. The profile is still good, and I give it here as a grid reference.

10   *NM 853 369 Nesøya.* – There is no exact locality for this bulk collection, which was donated by Dr. D. Worsley. The only possible place given on the geological map (Holtedahl & Dons 1977) is, however, a small hill on the south side of the island. Here the boundary between the Ordovician and Silurian is shown with only a small sliver of Silurian rocks.

11   NM *871 376 Brønnøya.* – Spot samples were collected only from the northeastern beach on the island. The basal meters of the Myren Member are exposed in a narrow syncline between Ordovician rocks. The exposure is on a narrow strait and continues on the other side of the strait on Ostøya in Bærum County.

12   *NM 874 376 Ostøya.* – Spot samples were collected at the base of the Myren Member (Solvik Formation) on the southwestern shore in the continuation of the same syncline found on Brønnøya. The rocks are very fossiliferous.

13   *NM 846 403 Kampebråten.* – Although not very well exposed, the entire 99 m thick Rytteråker Formation is followed by the 61 m of the Vik Formation (as defined by Möller 1989). The section is continuous but overgrown and at places strongly metamorphosed. It is a cliff section along a footpath leading from the Jongsåsveien to Jong school. Two very large bulk samples were taken in shale interbeds in the uppermost Rytteråker Formation (formerly one on each side of the boundary between the Rytteråker and Vik formations). Four bulk samples were also taken farther up into the Vik Formation through the first 22 m where fossils in shale were visible.

14   *NM 847 404 Jongsåsveien.* – This is a low hill confined on its three sides by the highway E 68, Jongsåsveien and a sidetrack to the railroad. The spot sampling was done in the hill on the north side of the road Jongsåsveien, where the uppermost 15–16 m of the Leangen Member (Solvik Formation) is succeeded by the Rytteråker Formation, which may be viewed on the railroad track.

15   *NM 846 413 Bærum nursing home.* – A few spot samples were taken in a short road cutting on the north side of Sykehusveien in front of the nursing home. The rocks were from the Leangen Member, Solvik Formation.

16   *NM 863 428 Christian Skredsviks vei.* – This is a small wooded outcrop on the west side of the road. Extensive spot sampling was done in shale intervals from the upper parts of the Rytteråker Formation *sensu* Möller (1989).

The city of Oslo

17   *NM 979 376 southwest coast of Malmøya.* – Coastal cliffs on the west to southwest coast of Malmøya expose the upper 60 m of the Myren Member, the entire Padda Member, both Solvik Formation, and the Rytteråker Formation. The section is exposed in an anticline, with possibilities for collection both on the west and south side of the island, but bulk samples and a few spot samples were collected on the southwest to west side.

18   *NM 976 378 north to northwest side of Malmøykalven.* – Steep cliffs with the Vik Formation are exposed on the north and northwest side of the island. The section is continuous with the Rytteråker Formation, but internal faulting makes the exact thickness uncertain. Bulk samples and spot samples were taken mainly from thin shale interbeds.

19   *NM 974 376 – 977 379 south side of Malmøykalven.* – A few spot samples were taken from this long beach section which consists of the upper parts of the Solvik Formation succeeded by the Rytteråker and Vik formations.

20   *NM 980 380 west side of Malmøya.* – Beach section and cliff section where the Vik Formation is exposed succeeding the Rytteråker Formation and followed by the Skinnerbukta Formation. Only a few spot samples from the boundary between the Vik and the Rytteråker formations were taken.

Measurements and statistics

The measurements were made using an eyepiece scale in a Zeiss microscope or, in very large specimens, a calliper. The measurements were made on moulds or, in a few instances, the actual shell; rubber casts were excluded. The sagittal length (s.l.), maximum width (m.w.), and hinge width (h.w.), all normally projected, were taken on all specimens where possible. In some instances maximum depth (m.d.) was also included. Measurements are given to the nearest half millimeter.

The measurements were treated in the statistical program 'Statview' (Version 4.02, Abacus Concepts, Inc. 1992–1993) on a Macintosh desk-top computer. Basic statistical computations and the variance–covariance matrix are given in Tables 3–53. This was only relevant when the number of measurements were adequate, e.g., larger than 3. Otherwise the raw numbers are given. The same program was used to create scattergrams with regression lines.

Geographic relationships of the fauna

Parts of the fauna found around the Ordovician–Silurian boundary are distinctly different from the remainder of the Llandovery fauna. The two faunas are therefore treated separately. I will first deal with the main brachiopod fauna found throughout the Llandovery section and then consider the boundary fauna components.

Comparison between Llandovery faunas

There is a very close connection between the brachiopods of the Oslo Region and those of the Welsh type district. This relationship ranges from the level of faunal associations down to species level. Contrary to Cocks *et al.* (1984, p. 153), I found (Baarli 1987) that the earliest Llandovery brachiopod fauna in Norway was organized in well-developed associations. Inspection of the faunal lists given in Cocks *et al.* (1984) shows that although the Early Llandovery collections are small, the brachiopods from the Crychan and Trefawr formations fall comfortably within the associations defined from Norway. The only exception are the *Cryptothyrella*-dominated associations from earliest Llandovery of the Bronydd Formation in Wales. This type of association is not found in Norway, probably because of differences in depth.

Furthermore, on a species level, with the exception of the Norwegian boundary fauna, all but some rare and new species (e.g., *Resserella matutina, Marklandella markesi, Salopina pumila, Hesperorthis hillistensis, Skenidioides hymiri, S. worsleyi, Eostropheodonta delicata, Eocymostrophia balderi,* and *Fardenia oblectator*), are found both in Norway and in Wales or southern England. Of these, the two *Skenidioides* species defined in Norway might be included in *Skenidioides* sp. (?*woodlandiensis* Davidson) of Temple (1987). Likewise,

the common *Isorthis prima,* found in all the Norwegian associations, may be included in '*Resserella*' sp. Temple (1987) listed '*Resserella*' sp. and *Isorthis prima* from Wales as synonyms of *Mendacella mullochiensis* [Davidson]. There are, however, a few local differences: *Protatrypa malmoyensis* in the Norwegian *Stricklandia*-coral association is substituted by *Plectatrypa* sp., and *Meifodia* sp. is more common in Wales.

In the rest of Scandia and Baltica proper, including the Russian Platform, the early Llandovery sequences were either not preserved or preserved as graptolitic facies. In the areas with a stratigraphic break the early Llandovery strata consist of carbonate facies, and the fauna there is different from the Norwegian–Welsh faunas deposited in more mixed argillaceous–carbonate facies. Exceptions include representatives of the eurytopic stricklandiids and pentamerids.

A fauna showing resemblance to the Welsh–Norwegian faunas is described by Kul'kov & Severgerina (1990) from Gornyj Altaj in Kazakhstan. The fossils are found in sandstones at the first Silurian fossiliferous level about 50–60m above an Ordovician fauna. Although the horizon is very poor in fossils, which limited the size of available collections, all the species reported are of a Welsh–Norwegian type. Also another impoverished fauna from Balqash Kol in eastern Kazakhstan have parts of the same fauna-assemblage (T.L. Modzalevskaja, personal communication, 1993).

In Laurentia, most of the Llandovery sequences are again represented by shallow carbonate facies. Exceptions are found in the Southern Midcontinent, where the northern parts present a deeper and more varied carbonate facies with richer brachiopod faunas than those of the interior of the craton. On a species level, however the brachiopods are distinctly different from the Norwegian fauna and even the Baltic carbonate fauna. The southeastern flank of the Southern Midcontinent are generally represented by coarse arenaceous and extremely shallow facies with a relatively poor fauna (Baarli *et al.* 1992). One exception is the Llandovery Red Mountain Formation at Ringgold in Georgia, where there is a relatively rich brachiopod fauna found in shales. The bulk of this fauna is of Benthic Assemblage 3; e.g., low diversity *Cryptothyrella* associations in lower Llandovery rocks, and later *Eocoelia* associations, but there is also a richer, deeper-water fauna. A brief investigation showed that the fauna there was related to those in Norway and Wales on a generic level. Most of the species, however, of both this and the *Cryptothyrella* associations seemed to be different from those of the Norwegian–Welsh fauna.

Geographic separation of the Llandovery faunas

This comparison leaves no doubt that there was a close connection between the part of Baltica where the Oslo Region was situated and Welsh Avalonia. The situation was much like that pictured by Cocks & Fortey (1982) and Fortey & Cocks (1992), with Avalonia docked to Baltica during the late Ordovician to early Silurian. A collision between Avalonia and Baltica in late Ordovician time is supported by tectonic activity both in Avalonia (Fortey & Cocks 1992) and Baltica (Baarli 1990b). This activity included development of a foreland basin with a peripheral bulge (Baarli 1990b) in the Oslo Region during the Llandovery. The nappes seemed to advance from the present west-north-west. A rotating Baltica (Torsvik *et al.* 1990) might account for the discrepant direction of these nappe-movements.

The connection with Kazakhstan and Gornyj Altaj is somewhat more complicated. Şengör *et al.* (1993) proposed an interesting model for the development of the Altaid orogeny. The two areas, previously mentioned, would be situated on a long island arc (the Kipchak arc) fastened to an inverted Siberian continent. The arc extended towards Baltica. They place Gornyj Altaj in the temperate zone, near the north side of Siberia in the Silurian. By comparison Baltica straddled the equator. It is hard to imagine the closely related Norwegian–Welsh and Altaid faunas living in different climatic zones and with no open sea-way between them. Rong & Harper (1988) added strength to this doubt. They claimed convincingly that different Upper Ordovician *Hirnantia* associations were related to different facies and climatic zones. There are similarities between the Edgewood fauna of the American midcontinent, the faunas of the Oslo Region, and the Gornyj Altaj Region. Rong & Harper (1988) therefore placed the three in the same climatic zone. Obviously the model of Şengör *et al.* (1993) requires some adjustment.

Baltica is thought to have collided with Laurentia in late Silurian, but already during Llandovery they shared some eurytopic and cosmopolitan species. The Southern Midcontinent represented an embayment in the Laurentian craton with the Taconic mountain chain on the east flank (Witzke 1990). The Taconic mountain chain and possibly also mountains developed in a collision between Baltica and Avalonia, might have been a considerable physical barrier. The separation between the platform faunas of the Baltic and the Laurentian cratons and between the more marginal faunas in tectonically active areas, were still large enough to warrant markedly different local species development in comparable facies and tropical position for both.

The boundary fauna

In the first 40 m of the Solvik Formation in Norway (e.g., near the Ordovician–Silurian boundary) there was a faunal component of Ordovician relation (Baarli & Harper 1986). A few of these are also found in this study: *Orthokopis idunnae* sp. nov., ?*Plectorthis* sp., *Skenidioides scoliodus*, and

Sowerbyella sp. *Orthokopis* seems to be the last representative of the subfamily Orthinae. If ?*Plectorthis* indeed belong to that genus, it is its youngest representative. *Skenidioides scoliodus* is found straddling the boundary in Britain and is therefore one of the few species that transcend the boundary.

The Norwegian boundary fauna has only a few taxa in common with those in Britain and none in the Altaids. Baarli (1988a) found that some of the enteletaceans were similar to those found in the uppermost Ordovician, deeper-water Králův Dvůr Formation of Bohemia. However, the present taxonomic treatment did not reveal any more taxa in common with that Bohemian fauna. There are some similarities with the deeper water *Aegironetes–Aulacopleura* community (upper Aeronian) in the Prague Basin (Havlíček & Štorch 1990). However, none of the dominant fauna-elements are shared in common. This is probably due to the differences in age and lithology, the Bohemian Želkovice Formation consisting predominantly of basalt tuffs and tuffaceous limestone compared with the shales of the lower Solvik Formation. It is worth noting that there is less connection between the lowermost fauna in the Solvik Formation and the uppermost faunas of the immediately underlying Langøyene Sandstone Formation (Cocks 1982; Brenchley & Cocks 1982) than between the faunas of the Solvik Formation and those mentioned from Bohemia. Some of these Norwegian, uppermost Ordovician, brachiopod associations show a striking resemblance to the North American Edgewood Fauna (Amsden 1974), as already mentioned, and also to the Kolyma fauna from Russia (Oradovskaya *in* Koren *et al.* 1983). However, all three faunas seem to have been connected to a tropical shallow-water environment, with tidal channels and oolite banks (Rong & Harper 1988). Brachiopods of the Solvik Formation lived in a deeper-water, offshore setting such as that evident for the Králův Dvůr and the Zelkovice formations of Bohemia. Environmental conditions seem therefore to be of critical importance. One would expect representatives of the offshore and marginal faunas to have the best chances to survive the major drop and subsequent rise in sealevel recorded close to the boundary. A tectonically active setting like the Bohemian basin and the basin of the Oslo Region could help counteract the eustatic changes, ensuring the short survival of deeper-water Ordovician relicts into the Silurian Period.

Systematic paleontology

This treatment of the superfamilies Orthacea and Davidsoniacea follows the systematic classificaton of Moore (1965). The superfamilies Plectambonitacea and Strophomenoidea are classified according to Cocks & Rong (1989) and Rong & Cocks (1994), respectively.

Class Calciata Popov, Bassett, Holmer & Laurie, 1993

Subclass Articulata Huxley, 1869

Order Orthida Schuchert & Cooper, 1932

Suborder Orthidina Schuchert & Cooper, 1932

Superfamily Orthacea Woodward, 1852

Family Orthidae Woodward, 1852

Subfamily Orthinae, Woodward, 1852

Genus *Orthokopis* gen. nov.

Type species. – *Orthokopis idunnae* sp. nov.

Derivation of name. – Greek *ortho-*, straight, and *kopis*, knife. The name is given in recognition of the blade-like brachiophores. The gender is feminine.

Diagnosis. – Ventribiconvex valves with costellate ribbing, often a pedicle fold, and brachial sulcus. The ventral muscle field is short, subpentagonal to oval, and elevated on a platform. The brachiophores are low, bladelike and unsupported, while the cardinal process is ridgelike and may thicken anteriorly. The dorsal muscle field is quadripartite and impressed.

Discussion. – This genus fits the diagnosis of the family Orthidae in all characters but its slightly bladelike brachiophores. However, according to Potter (1990), this trait is also seen in some species of *Orthambonites* Pander, 1830, and in *Diochthofera* Potter, 1990. Harper (1984) adds that bladelike brachiophores seem to be typical of the youngest species of *Orthambonites*, of Upper Ordovician age.

Diochthofera, known only from Ashgill strata of the eastern Klamath Mountains, Northern California, is possibly closest to *Orthokopis* in morphology and is of the same age. *Diochthofera* has a sulcate pedicle valve in contrast to the fold seen in *Orthokopis*, an elevated notothyrial platform, and pronounced, raised, and segmented edges around the dorsal muscle field, traits that are lacking in *Orthokopis*.

Harper (1984) pointed out that *Orthambonites* Pander, 1830, contained a large number of species, creating a fairly heterogenous group. He found that 'typical' *Orthambonites* possessed a costate ornament with fine ribs or capillae developed within the interspaces and brachiophores, which were massive and knob-like. Both of these characters differentiate it from *Orthokopis*. Harper (1984) proposed to limit the definition of *Orthambonites* to costate forms and exclude the costellate species. *Orthokopis* would therefore not fit within *Orthambonites*. Possibly, the costellate forms *Orthambonites*

ardmillanensis (Reed, 1917), *O. divaricatus* Cooper, 1956, *O. multicostellus* Cooper, 1956, and *O. tenuicostatus* Cooper, 1956, all of Middle Ordovician age, may be incorporated in *Orthokopis*.

Orthostrophia Hall, 1883, is another similar genus, but for its slightly converging brachiophore bases (which led Havlíček 1977 to remove it from the family Orthidae to the Plectorthidae) and well-developed notothyrial ridges.

Orthokopis idunnae sp. nov.
Pl. 1:1–3, 5–7, 9–11, 14, 15

Holotype. – An internal mould of a brachial valve (PMO 128.153, Pl. 1:1–2) from 9 m above the base of the Solvik Formation, Konglungø, Asker.

Derivation of name. – Named after the Norse goddess Idunn.

Material. – PMO 128.096, 128.101, 128.104, 128.124, 128.153, 128.197–128.201, 128.208: internal moulds of 2 brachial and 5 pedicle valves and 4 external moulds of valves, from the basal Myren Member of the Solvik Formation at Nesøya, Spirodden, Vakås, and Konglungø.

Diagnosis. – The valves are transversely subcircular to subrectangular in outline with moderately coarse costellate to fascicostellate ribbing. Small teeth are supported by strong dental plates. The muscle field is elevated on a callosity and has a rounded outline. The brachiophores are low and unsupported with faint notothyrial ridges. The muscle field is deeply impressed with the broader posterior scars separated by an oblique ridge from the anterior muscle scars.

Description. – Exterior: Transversely subcircular to subrectangular valves are $^7/_{10}$ as long as wide (Table 2), with ventribiconvex to planoconvex profiles. The brachial valve is moderately to slightly convex with a well-developed, broad sulcus originating near the umbo. The pedicle valve is moderately convex, deeper than the brachial valve, and has a fold.

Table 2. *Orthokopis idunnae* gen. et sp. nov., measurements and statistics.

Ventral valves	s.l.	m.w.	h.w.
Mean	5.5	8.33	8.5
Std. dev.	1.29	2.08	0.71
Count	4	3	2
Minimum	4.00	6.00	8.00
Maximum	7.00	10.00	9.00
Variance–	1.67	2.83	1.00
covariance		4.33	0.50
matrix			0.50
Dorsal valves	s.l.	m.w.	h.w.
PMO 128.153 (holotype)	12.00	12.00	12.00
PMO 128.096	10.00	14.00	–

The hinge line is straight and about $9/10$ of the maximum width of the shell. The maximum width occurs close to the valve mid-length, and the cardinal angles are rectangular. The lateral margins are slightly rounded. The anterior margin is evenly rounded and sulcate.

A plane, apsacline and fairly high pedicle valve exhibits a triangular delthyrium with an angle of 70–80°. The dorsal interarea is plane, anacline, and about $2/3$ as long as the ventral interarea. An open notothyrium is occupied medially by the cardinal process. The radial ornament is moderately coarse, costellate to fascicostellate. The costella are low, rounded triangular in cross-section, with broad rounded interspaces. New ribs are introduced largely by implantation. There are about 20 ribs per 10 mm at 5 mm distance from umbo.

Interior of pedicle valve: The delthyrial chamber is relatively shallow without pedicle callist. The small, blunt teeth have a triangular cross-section and shallow crural fossettes. The teeth are supported by strong but short dental plates, which are inclined towards the center of the delthyrial chamber. At their bases the dental plates are very short, sharply curved edges bounding a raised muscle field. The muscle fields are obscure but seem to consist of two relatively narrow, triangular diductor scars bordering an anteriorly protruding adductor scar. The width of the adductor scars is slightly more than the width of one diductor scar. The entire muscle scar has an oblong to rounded shape and is confined to the posterior $1/4$ of the shell. The mantle canal system is feebly impressed.

Interior of brachial valve: The notothyrial platform is weakly developed or lacking. The brachiophores are low and blade-like. They are fused to the valve floor and the notothyrial walls for most of their length; they diverge at about 90°. The cardinal process consists of a strong upright ridge, which thickens sharply anteriorly in most of the specimens. Very low, thin ridges may arise at the mid-length of the cardinal process and run parallel with the brachiophores. The muscle field is well impressed and consists of posterior scars that are broader than the anterior scars. They are separated by anterolaterally oblique ridges and a broad median area. The muscle fields occupy $1/3$ of the width and length of the valve. The mantel canal system is possibly lemniscate.

Remarks. – The scarce and fragmented material at hand made it questionable to erect a new taxon. However, the internal and external features of both valves are known, and it was then impossible to place the material comfortably within any existing genus or species. *Orthokopis idunnae* extends the age limits of the subfamily Orthinae up to the Ordovician–Silurian boundary and are thus the youngest species and genus of the Orthinae. Table 2 gives the measurements and statistics for *Orthokopis idunnae*. It is a relatively small species. Growth is possibly isometric, although the size range and number of specimens are too small to allow this to be demonstrated.

Family Dolerorthidae Öpik, 1934

Subfamily Dolerorthinae Öpik, 1934

Genus *Dolerorthis* Schuchert & Cooper, 1931

Type species. – *Orthis interplicata* Foerste, 1909, p. 76, from the Osgood Formation (Telychian) of Indiana, USA. By original designation of Schuchert & Cooper (1931, p. 244).

Dolerorthis sowerbyiana (Davidson, 1869)

Pl. 1:4, 8, 12–13, 16, 19, 22, 25

Synonymy. – ☐1839 *Spirifer plicatus* – J. de C. Sowerby *in* Murchison, p. 638, Pl. 21:6 [*nom. dub.*]. ☐1839 *Orthis radians* – J. de C. Sowerby *in* Murchison, p. 639, Pl. 22:11 [*nom. dub.*]. ☐1869 *Orthis Sowerbyiana* sp. nov. – Davidson [*pars*], p. 247, Pl. 35:27, 28, 30, 31 [*non* fig. 29]. ☐1917 *Orthis (Plectorthis) rustica* var. nov. *paucicostata* – Reed, p. 833, Pl. 6:9–12. ☐1970 *Dolerorthis sowerbyiana* (Davidson, 1869) – Temple, p. 13, Pl. 2:1–11. ☐1978 *Dolerorthis plicata* (J. de C. Sowerby, 1839) – Cocks, p. 43. ☐1982 *Dolerorthis plicata* (J. de C. Sowerby, 1839) – Thomsen & Baarli, Pl. 1:1. ☐1983 *Dolerorthis* sp. – Lockley, p. 94, Fig. 2 (1, 2). ☐1984 *Dolerorthis sowerbyiana* – Temple *in* Cocks *et al.*, pp. 150, 152–154. ☐1986 *Dolerorthis* sp. – Baarli & Harper, Pl. 1g and k, *non* f, h and l. ☐1987 *Dolerorthis sowerbyiana* (Davidson, 1869) – Temple, pp. 30–32, Pl. 1:11–18.

Lectotype. – Designated by Whittard & Barker 1950, p. 564. Internal and external moulds of dorsal valve, GSM 11606 and 11605, from Gasworks Mudstone [Haverford Mudstone Formation], Gasworks, Haverfordwest, Dyfed (Davidson 1869, Pl. 35:30; Whittard & Barker 1950, Pls. 1:11; 6:10).

Material. – PMO 103.516, 128.106–128.108, 128.113, 128.114, 128.116–128.118, 128.123, 128.133, 128.203, 128.204, 128.214, 128.221, 128.238, 130.893, 130.902–130.904: internal moulds of 11 pedicle and 8 brachial valves and external moulds of 2 valves, from the lower parts of the Myren Member, Solvik Formation, at Spirodden, Konglungø, Vakås, Nesøya and Hvalsodden.

Description. – Exterior: The valve is transversely subrectangular, $1/2$–$3/5$ as long as wide. The brachial valve is moderately and uniformly convex. The pedicle valve is variably developed from slightly convex to convexo-concave with the highest point near the posterior margin; it flattens out anteriorly and laterally. The valves lack sulcus or fold. The hinge line is straight. The lateral margins are gently rounded and continue into evenly and slightly rounded anterior margin. The commissure is crenulated and rectimarginate. The ventral interarea is relatively high, plane and apsacline. The

Table 3. Dolerorthis sowerbyiana (Davidson, 1869), measurements and statistics of the ventral valves.

	s.l.	m.w.	h.w.
Mean	10.00	16.70	17.44
Std. dev.	5.01	8.86	8.14
Count	8	10	9
Minimum	4.00	6.00	6.00
Maximum	17.00	31.00	28.00
Variance–	25.14	51.47	40.60
covariance		78.46	70.24
matrix			66.28

Table 4. Dolerorthis sowerbyiana (Davidson, 1869), statistics and measurements of the dorsal valves.

	s.l.	m.w.	h.w.
Mean	5.83	9.67	9.50
Std. dev.	2.48	4.84	4.93
Count	6	6	4
Minimum	3.00	4.00	4.00
Maximum	10.00	18.00	15.00
Variance–	6.17	11.73	14.50
covariance		23.47	29.83
matrix			24.33

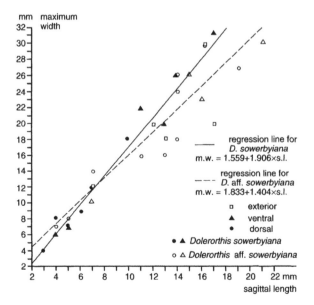

Fig. 3. Scattergram with regression lines for the early Rhuddanian population of *Dolerorthis sowerbyiana* (Davidson, 1869) and the early Aeronian population of *D.* aff. *sowerbyiensis* (Davidson, 1869), showing a close overlap in the saggial-length to maximum-width ratio throughout the different growth stages.

delthyrium is open, triangular with delthyrial angles of 50–60°. The dorsal interarea is plane and low. The notothyrium is open and occupied by the cardinal process medially. The ornamentation is radial with 12–15 ribs per 10 mm at the 10 mm growth stage. The costae are subangular in cross section with flat interspaces, broader than the costellae. Costellae arise with infrequent branching occurring in a position starting 5 mm from the umbo, in some instances with simultaneous branching to the right and left. Fine longitudinal striations or threads and strong fila are present in the interspaces all over the valve.

Interior of pedicle valve: The delthyrial chamber is deep. The teeth are strong and supported basally with strong, relatively short dental plates which are fused with the lateral walls of the delthyrium. The plates continue anteromedially as low, weak, and curved dental ridges, converging towards each other, but not enclosing the muscle field. The muscle field is faintly impressed and equally broad as long, occupying $\frac{1}{3}$–$\frac{2}{5}$ of the total length. The adductor scars are lanceolate and seldom impressed. The diductor scars are triangular and transversely striated. The mantle canal system may be discerned on one specimen as saccate with closely spaced vascula media. A thin, low median ridge may extend from the margin of the muscle field to separate the two mantle canal areas.

Interior of brachial valve: The cardinalia are raised on a low notothyrial platform that extends forward as a very short longitudinal medial ridge anterior between the muscle scars. The brachiophores are blade-like and short and supported for almost their entire length by anterolaterally directed brachiophore bases ankylosed onto the notothyrial platform. The sockets are shallow and supported by socket pads. The cardinal process is a very thin, low ridge with highest point medially and extending to the edge of the notothyrial platform. The muscle field is very weakly impressed.

Remarks. – *Dolerorthis sowerbyiana* is a medium-sized species that shows a wide variation in size (Tables 3, 4 and Fig. 3). However, the shape seems to be relatively size-independent and shows little variation for each growth stage (Fig. 3). *D. sowerbyiana,* as described above, occurs in the lower parts of the Rhuddanian together with ?*Hesperorthis gwalia.* The two are very close in internal characters. Like Temple (1987), I found that the main differences in the pedicle valve, lies in a markedly different ribbing. Differences in the brachial valve will be discussed under ?*H. gwalia.*

Dolerorthis aff. *sowerbyiana* (Davidson, 1869)

Pls. 1:17–18, 20–21, 23–24; 2:1–2, 4, 5

Synonymy. – □1982 *Dolerorthis plicata* (J. de C. Sowerby, 1839) – Thomsen & Baarli, p. 72, Pl. 1:1.

Material. – PMO 103.485, 105.196 (2 specimens), 128.138–128.148, 128.154, 128.155, 128.156 (2 specimens), 128.157, 128.158 (3 specimens), 128.159, 128.160, 130.892, 136.006: Internal moulds of 8 pedicle and 13 brachial valves and 5 external moulds, from the Leangen Member of the Solvik Formation at Skytterveien in Asker and Jongsåsveien in Sandvika. The species is also relatively common in the Spirodden Member, Solvik Formation, at Spirodden and Vettrebukta, Asker.

Remarks. – The populations of *D. sowerbyiana* found in the lowermost Rhuddanian and those in the lower half of the Aeronian are different as follows: The Aeronian material is nearly always pronouncedly mucronate even in juvenile material, in contrast to to the Rhuddanian populations where alae are not observed. The Aeronian material also has rounded ribs with interspaces generally narrower than the ribs. The rib density is 10 costella per 10 mm at the 10 mm growth stage. The sparse material of Rhuddanian age has angular ribs with wide, flat interspaces and a rib density of 12–15 costella per 10 mm at the 10 mm growth stage.

The Aeronian material differs from the Rhuddanian material also in internal characters. The pedicle muscle scars are much better impressed, broader, and larger relative to overall valve size, and the muscle-bounding ridges are stronger in the former while the pedicle umbo is more erect in the latter. In the brachial valve, the Aeronian material has a higher interarea leading to relatively deeper notothyrial chamber; the brachiophores also extend farther outside the notothyrium than those in the Rhuddanian material.

Temple (1987) mentioned that the detail of ribbing and surface sculpture of *D. sowerbyiana* varies and may be used for discrimination of samples. The Rhuddanian valves from Norway are close to the ribbing shown in Pl. 1:12 of Temple (1987) from the Llandovery in Wales, although the interspaces in the Norwegian material is even broader relative to the ribs than in the Welsh material. The Aeronian material, however, is like the material figured in Temple (1970) for *D. sowerbyiana* from Mathrafal, Montgomeryshire. Another difference is that Temple's material has a sulcus in the brachial valve that dies out anteriorly. The Norwegian Rhuddanian material seems to lack this feature, although the available brachial valves are very small. In the Aeronian-aged material it is developed in the smaller specimens. Internally the Rhuddanian material agrees best both with the material from Wales and from Montgomeryshire.

Fig. 3 shows that although the Rhuddanian material tends to be smaller than that from the Aeronian, there are no clear differences in plots of size or valve shape in the two populations. The slightly low covariance between sagittal length and maximum width (Tables 5, 6) for the Aeronian material is explained by the great variation in shape at each growth stage, as seen in Fig. 3.

From the above discussion, clearly the differences in the two Norwegian samples may be topical; perhaps related to the more proximal setting and higher energy levels that existed for the Aeronian populations compared with the Rhuddanian material (Baarli 1985). The differences in morphology may not be large enough to warrant placement in two different species. The Rhuddanian material is, however, closest to the British material.

Table 5. Dolerorthis aff. *sowerbyiana* (Davidson, 1869), measurements and statistics of the ventral valves.

	s.l.	m.w.	h.w.
Mean	15.67	22.33	22.33
Std. dev.	4.97	6.77	6.77
Count	6	6	6
Minimum	7.00	10.00	10.00
Maximum	21.00	30.00	30.00
Variance–	24.67	36.25	36.25
covariance	36.25	45.87	45.87
matrix	36.25	45.87	45.87

Table 6. Dolerorthis aff. *sowerbyiana* (Davidson, 1869), measurements and statistics of the dorsal valve.

	s.l.	m.w.	h.w.
Mean	11.70	20.67	18.10
Std. dev.	4.90	8.66	7.78
Count	10	12	10
Minimum	4.00	7.00	7.00
Maximum	19.00	32.00	30.00
Variance–	24.01	35.42	32.31
covariance		74.97	63.12
matrix			60.54

Dolerorthis aff. *psygma* Lamont & Gilbert, 1945

Pl. 2:7–8, 10–11, 13–14

Synonymy. – ☐1982 *Dolerorthis* aff. *psygma* – Cocks & Baarli, pp. 80 and 82.

Material. – PMO 43.385, 43.387 (two specimens), 43.388–43.394, 43.444, 43.445, 43.447: 5 brachial valves and one internal mould of a brachial valve, 3 pedicle valves and one internal mould of a pedicle valve, and 1 complete shell from the upper parts of the Rytteråker Formation and the entire Vik Formation at Malmøya.

Description. – Exterior: The valves are subequally biconvex to dorsibiconvex. The pedicle valve tends to flatten anteriorly and laterally. Generally the brachial valve is non-sulcate. The shell is transversely subcircular in outline and about $4/5$ as long as wide. A straight hinge line is $4/5$ of maximum width; the cardinal angles are rounded. The maximum width lies between $1/3$ and $1/2$ length from the umbo. The margins are

Table 7. Dolerorthis aff. *psygma* (Lamont & Gilbert, 1945), measurements of valves.

Ventral valves	s.l.	m.w.	h.w.
PMO 43.385 external valve	23.00	–	–
PMO 43.387 external valve	26.00	30.00	–
PMO 43.445 internal mould	19.00	23.00	20.00

Dorsal valves	s.l.	m.w.	h.w.
PMO 43.387 external valve	–	30.00	–
PMO 43.388 external valve	26.00	30.00	24.00
PMO 43.389 internal mould	–	37.00	30.00
PMO 43.390 external valve	24.00	29.00	–
PMO 43.391 internal mould	24.00	29.00	22.00
PMO 43.393 external valve	–	31.00	–
PMO 43.444 external valve	12.00	20.00	–

evenly rounded with a crenulated commissure that is rectimarginate anteriorly.

The ventral beak is erect. The ventral interarea is apsacline and about 3 times as long as the anacline dorsal interarea. The delthyrium is open with variable delthyrial angles, 30–60°. An open notothyrium is occupied medially by the posterior parts of the cardinal process.

The radial ornament is coarsely costellate with less than 2 costella per millimeter at the 10 mm growth stage of the brachial valve. The ribs are subangular with wide, rounded interspaces and non-branching outside the umbo area. Strong fila, which are especially visible between the ribs, are developed all over the valve.

Interior of pedicle valve: The delthyrial chamber is narrow and shallow with a flat floor. The teeth are weak, subtriangular and supported basally by short, anterolaterally diverging dental plates. Short, weakly developed dental ridges curve slightly anteromedially before they die out. The muscle field is not deeply impressed on the one mould available. Width of the muscle field equals length. The muscle field occupies only $\frac{1}{4}$ of the total valve length and $\frac{1}{4}$–$\frac{1}{5}$ of the total width. The surface of the valve, with the exception of the muscle area, is impressed by ribs.

Interior of brachial valve: The cardinalia are situated on a well-developed notothyrial platform that continues in a steeply descending, short, and rounded median ridge. The ridge narrows and tapers out at about $\frac{1}{4}$ valve length from umbo. The cardinal process consists of a simple, thin, linear ridge extending across the notothyrial platform. It expands slightly in width and greatly in height in its anterior portions. The brachiophores are low, stout and subtriangular in cross-section with a convex median face, or rather a bulge, creating a faint ridge longitudinally on the brachiophores. The brachiophores are anterolaterally diverging and extending to the edge of the notothyrial platform. The sockets are very shallow and small, supported on thick socket pads. The muscle field is weakly impressed, consisting of very small

circular posterior fields occupying $\frac{1}{4}$–$\frac{1}{3}$ of the total width and separated by a wide expanse including the median ridge. The anterior fields are indistinct, but probably somewhat larger. The ribbing is impressed over the entire valve, except for the posterior muscle fields.

Remarks. – This material is too sparse for statistical treatment, but Table 7 shows good consistency in the size-range. The species is closely related to *D. psygma* in its style of ribbing, which distinguishes them from other European species. However, it differs from *D. psygma* in absence of a dorsal fold and in internal characters. These are obvious first and foremost in the pedicle valve, where in the Norwegian material the beak is more erect, the muscle scars much smaller, and more weakly impressed. The brachial valves are similar but the internal impressions of the ribs are not 'furrowed', e.g., show no median depressions marginally as described by Lamont & Gilbert (1945).

Genus *Schizonema* Foerste, 1909

Type species. – Subsequently designated by Foerste (1912, p. 139); *Herbertella* (*Schizonema*) *fissistrata* Foerste, 1909, p. 76; from the Silurian (Niagaran) of Indiana.

Schizonema subplicatum (Reed, 1917)

Pls. 2:3, 6, 9, 12, 15–18; 3:1–2

Synonymy. – □1917 *Orthis calligramma* Dalman var. nov. *subplicata* – Reed, p. 828, Pl. 5:10–15. □1949 *Orthis subplicata* var. nov. *varicosa* – Bancroft, p. 2, Pl. 1:1–4. □1949 *Orthis subplicata* var. nov. *neptuni* – Bancroft, p. 3, Pl. 1:7. □1949 *Strophomena antiquata* (J. de C. Sowerby) – Bancroft, p. 13, Pl. 1:8. □1951 *Schizoramma* cf. *subplicata* (Reed) – Williams, p. 90, Pl. 3:4–8, Text-fig. 2. □1970 *Schizonema* sp. – Temple, p. 17, Pl. 3:17–18. □1978 *Schizonema subplicatum* (Reed, 1917) – Cocks, p. 45. □1982 *Schizonema subplicatum* (Reed, 1917) – Thomsen & Baarli, Pl. 1:7. □1986 *Schizonema* (Reed) – Baarli & Harper, Pl. 1a–e. □1987 *Schizonema subplicatum* (Reed, 1917) – Temple, pp. 33–35, Pl. 2:5–11.

Type material. – Lectotype of *subplicatum*. Designated by Cocks (1978, p. 45). Internal mould of dorsal valve, B74483, from the Mulloch Hill Formation, Mulloch Hill Wood, Girvan, Strathclyde (Reed, 1917, Pls. 2:5; 5:11).

Material. – PMO 105.200–105.202, 107.590–107.592, 128.093–128.095, 128.097, 128.098, 128.109, 128.110, 130.895, 130.896, 130.900: internal moulds of 6 brachial and 6 pedicle valves and 4 external moulds, from the Myren Member of the Solvik Formation at Nesøya, Vakås, Brønnøya, and Konglungø.

Description. – Exterior: The brachial valve is flat to slightly concave with a weak median sulcus. The convex pedicle valve has the highest point close to the umbo. The outline is semi-circular, with length $^1/_1$–$^3/_4$ of width. The hinge line is straight and equal to or slightly shorter than maximum valve width. The cardinal angles are obtusely rounded to straight and continue anteriorly into evenly rounded lateral margins. The commissure is crenulated, and the anterior margin is recti-marginate. A high, plane, and apsacline ventral interarea has an open, triangular delthyrium with an angle of 60–70°. The dorsal interarea is narrow, flat, and anacline. The noto-thyrium is open and occupied medially by the cardinal process.

A coarse, costate to fascicostellate radial ornament counts 6 costae per 10 mm at the 10 mm growth stage in the brachial valve. The costae are rounded to triangular in cross-section and tend to be irregular and fascicostellate in mature speci-mens. Internal branches of the ribs appear at about 8 mm from umbo on most valves. Fila are well developed. The ribbing is impressed over most of the valve interior.

Interior of pedicle valve: The delthyrial chamber is deep with strong and blunt teeth. Short dental plates are inclined slightly towards the midline of the floor and fused with the lateral walls of the delthyrium. The dental plates continue anteriorly as low, short, and slightly curved ridges which fuse with the vascula media. The vascula media converge initially anteriorly and thereafter diverge. The muscle field is about as long as wide. The triangular diductor scars lie lateral to broadly lanceolate adductor scars that extend anterior of the diductor scars.

Interior of brachial valve: The cardinalia are raised on a high, broad platform that continues anteriorly in a broad descending ridge, reaching the mid-length of the valve. The brachiophores are triangular in outline. Notothyrial ridges running parallel to the brachiophores are well developed in most valves. Brachiophores are ankylosed to the noto-thyrial platform for most of their length and continue just slightly outside the platform. Shallow dental sockets are resting on socket pads. The cardinal process is a thin ridge widening anteriorly in most valves. The muscle field is weakly impressed.

Remarks. – Tables 8 and 9 show that *S. subplicatum* is a medium-sized species with a relatively good covariance be-tween sagittal length and maximum width, given the few valves present. This material closely resembles *S. subplicatum* as described by Williams (1951) and, more recently, by Temple (1987).

There are some doubts, however, whether this material should be described as *Dolerorthis* or *Schizonema*. Bassett (1970) reviewed the two genera and concluded that the only difference between the two lay in the presence of notothyrial ridges for *Schizonema* as opposed to a lack of them in *Doler-orthis*. This was in line with Boucot (1960), although he also stressed the lack of strong fila. However, the definition of *Dolerorthis* given by Williams & Wright (1965) notes that the cardinalia may have variably developed notothyrial ridges. Rudimentary notothyrial ridges are, for example, seen in

Table 8. Schizonema subplicatum (Reed, 1917), measurements and statistics of the ventral valve.

	s.l.	m.w.	h.w.
Mean	18.50	18.75	19.67
Std. dev.	2.89	4.99	5.69
Count	4	4	3
Minimum	15.00	15.00	15.00
Maximum	22.00	26.00	26.00
Variance–covariance matrix	8.33	10.17 24.92	10.17 32.33 32.33

Table 9. Schizonema subplicatum (Reed, 1917), measurements and statistics of the dorsal valve.

	s.l.	m.w.	h.w.
Mean	7.17	13.60	14.67
Std. dev.	5.92	7.02	6.11
Count	3	5	3
Minimum	3.50	6.00	8.00
Maximum	14.00	22.00	20.00
Variance–covariance matrix	35.08	43.08 49.30	0.00 31.33 37.33

Dolerorthis aff. *psygma* (Pl. 2:11). Thus the presence or ab-sence of notothyrial ridges is not a valid character for differ-entiation.

The type *Schizonema fissistriata* has a plane to gently con-vex brachial valve with a sulcus. The ribbing, especially in mature specimens, tends to be fascicostellate (Bassett 1970, Pl. 2:13b, 15a). All these characteristics are also seen in *S. subplicatum* (the best developed fascicostellate ribbing is shown in Pl. 3:2). I propose that fascicostellate ribbing in mature specimens is included in the definition. *S. hami* Amsden (1951) is the species pictured as *Schizonema* in *Treatise on Invertebrate Paleontology* (Williams & Wright 1965). Amsden (1968) reassigned this species to *Dolerorthis* because it lacked notothyrial ridges. It has the slight fasci-costellate ribbing, and the flat valve with brachial sulcus and would remain under *Schizonema* if the external ribbing and form were the main criteria used.

Subfamily *Hesperorthinae* Schuchert & Cooper, 1931

Genus *Hesperorthis* Schuchert & Cooper, 1931

Type species. – *Orthis tricenaria* Conrad, 1843, p. 333; from the Trenton Limestone (late Ordovician) of Mineral Point, Wisconsin, USA. By original designation of Schuchert & Cooper (1931, p. 244).

Hesperorthis hillistensis Rubel, 1962

Pl. 3:3–10, 12–13

Synonymy. – □1858 *Orthis Davidsoni* – Schmidt (*pars*), p. 214. □1928 *Orthis Davidsoni* Vern. – Teichert (*pars*). □1962 *Hesperorthis hillistensis* sp.n. – Rubel, pp. 84–86, Pl. 2:10–19, Pl. 3:1–4.

Type specimen. – Complete valve, BR 2836 from horizon G2H (lower to middle Llandovery) at Hilliste, northern Estonia.

Material. – PMO 109.889, 135.950–954, 135.957, 135.959, 135.961, 135.962, 135.972, 135.974, 135.976–983, 135.985, 135.988–991, 136.005: internal moulds of 14 pedicle and 8 brachial valves and external mould of 1 pedicle and 3 brachial valves, from near the base of the Vik Formation at Christian Skredsviks vei in Bærum and Kampebråten, Sandvika.

Description. – Exterior: The valves are planoconvex to ventribiconvex with a subcircular outline. The hinge line is straight with rounded cardinal angles and evenly rounded margins. There is a considerable variation in the ratio of maximum length to maximum width, which varies from $\frac{1}{2}$ to $\frac{1}{1}$. Some of the variation may be attributed to distortion. The hinge line is generally $\frac{9}{10}$ of maximum width, which occurs close to mid-length. The commissure is coarsely crenulated with a rectimarginate anterior part.

The ventral beak is erect and the interarea is high, apsacline and concave to a variable degree. Again the concave interarea may be due to distortion. The delthyrium is narrow with

angles of 20–30°. A small apical plate is preserved in some of the specimens. The dorsal interarea is very low with an open notothyrium.

The ornament is coarsely costate with 8 ribs per 10 mm at the 10 mm growth stage. This corresponds to a total of 18–21 ribs on the few valves where a count was possible. The costae are rounded although somewhat flattened on the top while the interspaces are slightly narrower and evenly rounded. Coarse fila and fine longitudinal threads are found in the interspaces.

Interior of pedicle valve: The delthyrial chamber is deep with small teeth supported by short, laterally convex dental plates. The dental plates extend anteriorly as short, thick dental ridges, which curve sharply towards the midline and may delimit the muscle field anteriorly. The muscle field consists of broad, triangular diductor scars, often transversely striated, and thin lanceolate adductor scars, which extend slightly anterior of the diductor scars. A thin, low median ridge sometimes extends from the anterior end of the muscle scar and separates the parallel vascula media. The periphery of the valve is strongly crenulated, and impressions of ribs may be seen over the whole valve outside the muscle field.

Interior of brachial valve: The brachiophores are short, thick blades to rods diverging at about 80°, and ankylosed directly to the valve floor. The sockets are very shallow. A variable cardinal process is mostly developed as a thin, low ridge, but may thicken anteriorly. A thickening of the valve floor produces a low ridge between the anterior ends of the brachiophores directly in front of the cardinal process. Posterior of the ridge in the notothyrial chamber there is a hollow on each side of the cardinal process. A faint, short median ridge may be seen midway down the valve. The muscle scars are not impressed, while the periphery of the valve is strongly crenulated.

Remarks. – Tables 10–11 give the measurements and statistics for *H. hillistensis.* It is a medium-sized to small brachiopod with an apparent size-independent shape. The low covariance between sagittal length and maximum width in the dorsal valve may be explained by a relatively large amount of shape variation of the dorsal valve, as seen in Fig. 4.

Owing to distortion of the shell material, the assignment to species is slightly questionable. The two closest species are *H. imbecilla* Rubel, 1962, and *H. hillistensis* Rubel, 1962. The difference between the two lies in the concave interarea of the former compared to a flat interarea of the latter. *H. hillistensis* has a higher, subpyramidal pedicle valve and a flatter brachial valve than *H. imbecilla.* The Norwegian material has a pedicle valve varying from very high to nearly flat, but the shape is clearly affected by tectonic distortion. It also has an apsacline concave interarea, but this may also be due to distortion. The interarea, however, is high, which supports affinity to *H. hillistensis,* as do the consistently flat brachial valve and the number of costae, which is in the low range of *H. hillistensis* but outside the range of *H. imbecilla.*

Table 10. Hesperorthis hillistensis (Rubel, 1962), measurements and statistics of the ventral valve.

	s.l.	m.w.	h.w.
Mean	12.20	14.50	14.71
Std. dev.	4.34	4.70	5.19
Count	10	10	7
Minimum	6.00	7.00	7.00
Maximum	19.00	22.00	22.00
Variance–	18.84	23.16	27.47
covariance	23.16	22.06	27.14
matrix	27.47	27.14	26.90

Table 11. Hesperorthis hillistensis (Rubel, 1962), measurements and statistics for dorsal valves.

	s.l.	m.w.	h.w.
Mean	9.64	13.27	13.75
Std. dev.	2.29	4.17	3.33
Count	11	11	9
Minimum	6.00	7.00	10.00
Maximum	14.00	22.00	19.00
Variance–	5.25	6.91	3.86
covariance		17.42	13.11
matrix			11.07

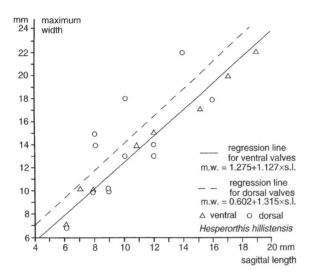

Fig. 4. Scattergram with regression lines for ventral and dorsal valves of *Hesperorthis hillistensis* Rubel, 1962, plotting sagittal length against maximum width.

?*Hesperorthis gwalia* (Bancroft, 1949)

Pl. 3:11, 14–20. Pl. 4:1–2

Synonymy. – □1949 *Orthis gwalia* sp. nov. – Bancroft, p. 3, Pl. 1:8, 9. □1951 *Giraldiella protensa filitexta* subsp. nov. – Williams [*pars*], p. 93, Pl. 3:13, 14 [*non* :15]. □1970 *Dolerorthis* sp. nov. – Temple, p. 15, Pl. 2:12–17. □1978 *Hesperorthis gwalia* (Bancroft 1949) – Cocks, p. 46. □1984 *Dolerorthis* sp. – Temple *in* Cocks *et al.*, p. 152. □1986 *Dolerorthis* sp. – Baarli & Harper, Pl. 1f, h, i, *non* g, k. □1987 *Dolerorthis gwalia* (Bancroft, 1949) – Temple, pp. 32–33, Pl. 1:19–25.

Lectotype. – Designated by Cocks 1978, p. 46. Internal mould of dorsal valve, SM A35687, from early Llandovery, 300 m ENE of Pen-y-rhiw, about 1.6 km ESE of Llanfair-ar-y-bryn, NE of Llandovery, Dyfed (approx. grid. ref. SN 824 3930); Bancroft 1949, Pl. 1:9, 19.

Material. – PMO 107.593, 107.595, 128.105, 128.107, 128.111–128.113, 128.115, 128.119, 128.121, 128.126, 128.129, 128.132, 128.212, 128.216, 128.217, 128.220, 128.221, 128.233 (two specimens), 130.894, 130.898, 130.923: internal moulds of 11 brachial and 9 pedicle valves and 3 external impressions, from the Myren Member of the Solvik Formation at Konglungø, Spirodden, and Vakås in Asker.

Description. – Exterior: The valves are ventribiconvex with a transverse to semi-circular outline, $^3/_5$–$^4/_5$ as long as wide. There is a shallow sulcus in the brachial valve originating near the umbo, which reaches the anterior margin in some specimens. The anterior margin is uniplicate to rectimarginate and crenulated, and the lateral margins are evenly rounded. The larger valves may rarely be slightly alate. Maximum width is at or immediately anterior of the hinge line

The ventral interarea is plane, apsacline and generally with an open delthyrium between 50° and 60° in angle. One specimen (PMO 128.107), however, seems to have an apical plate (Pl. 3:16). The dorsal interarea is anacline and somewhat higher than in *D. sowerbyiana*, which it occurs with. The notothyrium is open and occupied medially by a cardinal process. The ornamentation is radial with about 19 ribs per 10 mm at the 10 mm growth stage. The ribs are rounded in cross section and frequently branch by intercalation. They show a slight tendency to bunch together in a fascicostellate fashion (Pl. 3:18–19). There are strong fila between the ribs, giving the ribbing on the mould a beaded appearance.

Interior of pedicle valve: The delthyrial chamber is fairly deep. The teeth are strong and broadly triangular in cross section. They are supported basally by very short and relatively thin dental plates that are fused with the lateral walls of the delthyrium. The dental plates continue at their bases as short, low, curved ridges that are convergent onto the midline of the valve, bounding the muscle field anterolaterally. The muscle field is variably impressed, occupying $^1/_3$ of the total valve length. The diductor scars are triangular and often transversely marked, flanking the narrow, lanceolate adductor scars. A low median ridge may extend anteriorly from the muscle scars. The mantle system is obscure and often overprinted by ribbing.

Interior of brachial valve: The cardinalia are raised on a strong and relatively high notothyrial platform. The platform extends anteriorly as a broad, rounded ridge, reaching mid-valve length and sometimes continuing as a broad, shallow sulcus. The brachiophores are long, slender, and blade-like, extending beyond the short notothyrial platform. Their bases are diverging slightly laterally onto the notothyrial platform. One specimen (PMO 107.593) shows weakly developed notothyrial ridges. The sockets are shallow with grooves and lacking socket pads. The cardinal process is thin, long and extending to the edge of the notothyrial platform. The muscle field is quadripartite and small, with the anterior muscle field larger than the posterior. The fields are divided by oblique, low ridges running anterolaterally. The mantle canal system is feebly impressed.

Remarks. – This is a medium-sized species with a wide spread in size for the relatively sparse material present (Tables 12, 13). There seems to be consistent shape throughout the material.

The taxon occurs together with *D. sowerbyiana* (Davidson, 1869) and may be very difficult to distinguish from that species when no ribbing is visible, especially on small pedicle valves. The differences in the pedicle valve are that ?*H. gwalia* has thinner, shorter dental plates and more curved ridges, tending to enclose the muscle scar. *D. sowerbyiana* has a more pronounced umbo. The brachial valve in ?*H. gwalia* has long, straight, blade-like brachiophores much like *Hesperorthis*. This is in contrast to the shorter, triangular and slightly curved brachiophores in *D. sowerbyiana*. Other dif-

Table 12. ?*Hesperorthis gwalia* (Bancroft, 1949), measurements and statistics for the ventral valves.

	s.l.	m.w.	h.w.
Mean	11.80	18.20	17.80
Std. dev.	2.49	4.71	4.82
Count	5	5	5
Minimum	10.00	14.00	14.00
Maximum	16.00	26.00	26.00
Variance–	6.20	16.67	19.00
covariance		22.20	22.30
matrix			23.20

Table 13. ?*Hesperorthis gwalia* (Bancroft, 1949), measurements and statistics for the dorsal valves.

	s.l.	m.w.	h.w.
Mean	11.57	13.50	12.00
Std. dev.	5.09	4.07	4.51
Count	7	8	7
Minimum	6.00	10.00	7.00
Maximum	18.00	20.00	20.00
Variance–	25.95	21.40	26.30
covariance		16.57	18.67
matrix			20.33

ferences distinguishing the two are that ?*H. gwalia* often lacks socket pads, often has a well developed sulcus, and may have very fine notothyrial ridges in large specimens.

?*H. gwalia* has had a difficult taxonomic history being placed in different genera by all the authors who have discussed it. Temple (1970 and 1987) placed it in the genus *Dolerorthis* and subfamily Dolerorthinae, while Cocks (1978) placed it in *Hesperorthis* in the subfamily Hesperorthinae. The present study agrees with Cocks (1978) that ?*H. gwalia* is closer to *Hesperorthis,* since the cardinalia are very similar. This interpretation would be strengthened if the pedicle valve (PMO 107.128) indeed possesses an apical plate, but the mode of preservation makes this somewhat uncertain. Apical plates may easily be destroyed during preservation, as stated by Ross (1959) and Chiang (1972), among others, so the lack of apical plates in the rest of the material may be explained. However, the ribbing that was unknown to Cocks (1978) is costellate with slightly curved costella. Typical *Hesperorthis* has straight costate or paucicostate ornament, and Chiang (1972) proposed to restrict *Hesperorthis* to such species. *Lordorthis* Ross, 1959, shares the fashion of ribbing and the form of cardinalia with the present species. However, the biconvex shape in the present taxa disagrees with the pyramidal ventral valves in *Hesperorthis* and *Lordorthis* as well as with the flat to gently concave dorsal valve in the former and the resupinate dorsal valve in the latter. ?*H. gwalia* thus does not comfortably fall within any known genus of Hesperorthids.

Bassett & Cocks (1974) described two biconvex and multicostellate species of *Hesperorthis* [*H. gothlandica* (Schuchert & Cooper, 1932) and *H. martinssoni* Bassett & Cocks, 1974] from the Silurian of Gotland. Bassett & Cocks (1974) agreed with Chiang (1972) that multicostellate species now referred to *Hesperorthis* may best be separated from *Hesperorthis*, but at a subgeneric level. The present species, ?*H. gwalia*, is clearly closely related to *H. gothlandica* and *H. martinssoni*, and I follow Bassett & Cocks (1974) in keeping the name *Hesperorthis*. However, I add a question mark to indicate that this group of species ought to be separated from *Hesperorthis* when all the species of that genus have been reinvestigated. Since subgenera are usually not used within the Orthacea, a new genus name would probably be appropriate.

Subfamily Glyptorthinae Schuchert & Cooper, 1931

Genus *Eridorthis* Foerste, 1909

Type species. – Subsequently designated by Schuchert & LeVene (1929, p. 58): *Plectorthis (Eridorthis) nicklesi* Foerste, 1909, p. 222, from the Ordovician (Cincinnatian) of Kentucky, USA.

Eridorthis vidari sp. nov.
Pl. 4:3–11, 14

Synonymy. – □1970 Orthacea indet. – Temple [*pars*], p. 20 [?*non* Pl. 3:19]. □1987 *Eridorthis* sp. – Temple, pp. 35–36, Pl. 2:19–24.

Holotype. – PMO 128.188 exterior mould of a pedicle valve from the upper parts of the Leangen Member, Solvik Formation, Leangbukta, Asker.

Derivation of name. – Named after the Norse god Vidar.

Material. – PMO 103.475, 128.181–128.187, 128.188 (two specimens), 128.189–128.191, 128.193 (two specimens), 128.194 (two specimens): internal moulds of 5 pedicle and 8 brachial valves and 4 valve exteriors from the upper parts of the Leangen Member (Aeronian) of the Solvik Formation at Leangbukta and Skytterveien, Asker.

Diagnosis. – The form is biconvex, with well developed imbrication drawn out as short frills. A very shallow sulcus is developed only in mature pedicle valves. The brachial valve bears a narrow sulcus originating near the umbo but not reaching the anterior margin. A semi-circular ventral muscle scar is weakly impressed.

Description. – Exterior: The semi-circular valves are about $\frac{3}{4}$ as long as wide, with length possibly increasing anisometrically during ontogeny relative to width. The brachial valve is evenly and moderately convex and more convex than the pedicle valve. A fold is not well developed, but a narrow sulcus originates near the umbo in the brachial valve (Pl. 4:14, to the right) although not reaching the anterior margin. The largest pedicle valve displays a slight sulcus. The hinge line is straight and equal to $^9/_{10}$ of the maximum width. Rounded cardinal angles continue into evenly curved lateral and anterior margins. The commissure is rectimarginate. The ventral interarea is plane, low, and apsacline with open delthyrium with angles of 80–90°. The dorsal interarea is low, plane, and anacline, with open notothyrium occupied medially by the cardinal process.

Radial and strongly imbricated ornamentation is drawn out as frills with 7–8 costellae per 10 mm at the 10 mm growth stage of the brachial valve. Anteriorly, a few costellae may appear by insertion.

Interior of pedicle valve: The delthyrial chamber is shallow without pedicle callist. Blunt, triangular teeth are supported basally by short, relatively thin, and slightly divergent dental plates. The dental plates curve and continue anteriorly in low, thin ridges that enclose the muscle scars, giving the muscle field a semi-circular outline. The muscle scars are limited to the delthyrial chamber and occupy $^1/_3$ of the total valve length. Lanceolate adductor scars are about as wide as one of the triangular diductor scars but not elevated on a pad. The diductor scars may be marked transversely. The mantle canal system is unknown, owing to overprint by ribbing.

Interior of brachial valve: The cardinalia are raised on a high but narrow notothyrial platform that continues in a narrow rounded ridge, stretching $^3/_4$ of the valve length anteriorly. This ridge descends steeply from the notothyrial platform and continues as a low ridge corresponding to a shallow external sulcus that flattens out towards the anterior margin. Straight, relatively low brachiophores diverge at about 90°. They are fused to the notothyrial platform for most of their length. The sockets are deep and small. A strong, high, ridge-like cardinal process may thicken anteriorly. It reaches its maximum height at mid-length. The muscle scars are overprinted by ribs and the mantle system is unknown for the same reason.

Remarks. – Tables 14 and 15 give the measurements and statistics for the available material. There are no indications of allometry, although the number of specimens is inadequate for a reliable analysis.

Temple (1987) described *Eridorthis* without reference to a species because of inadequate material. The present material seems to be very similar to *Eridorthis* sp. of Temple (1987). The few differences observed are a coarser ribbing, weaker ventral plates and lower ventral interarea in the Norwegian material compared with those of the Welsh material. However, the Norwegian material consists of smaller specimens which may account for some of the differences. Generally the differences do not warrant two separate species.

Table 14. Eridorthis vidari sp. nov., measurements and statistics for the ventral valve.

	s.l.	m.w.	h.w.
Mean	8.67	11.00	8.75
Std. dev.	3.51	3.16	4.11
Count	3	4	4
Minimum	5.00	8.00	5.00
Maximum	12	15.00	14.00
Variance–	12.33	12.33	14.00
covariance		10.00	12.67
matrix			16.92
PMO 128.188 (holot.) ext. mould	10.00	14.00	12.00
PMO 128.188 (parat.) int. mould	5.00	8.00	6.00
PMO 128.193 (parat.) int. mould	9.00	12.00	10.00

Table 15. Eridorthis vidari sp. nov., measurements and statistics of the dorsal valves.

	s.l.	m.w.	h.w.
Mean	8.86	11.17	8.75
Std. dev.	2.91	3.25	1.89
Count	7	6	4
Minimum	5.00	7.00	6.00
Maximum	14.00	16.00	10.00
Variance–	8.48	10.23	4.50
covariance	10.23	10.57	4.92
matrix	4.50	4.92	3.58

In his review of the genus, Percival (1991) dismisses Temple's *Eridorthis* as atypical for the genus without specifying his criteria. It seems, however, that the ventral impressions of the muscle fields are different from most species, e.g., *E. vidari* lacks the well-developed, elevated median ridge and has generally smaller ventral muscle scars than normal. *Eridorthis* is relatively rare, although geographically widespread (Percival 1991), and except for *E. novadomis* Bassett (1972) from Wenlock strata in Wales, it is only known from the Ordovician. *E. vidari* fills this gap and at the same time widens the geographic range. *E. vidari* is close to *E. novadomis,* but the former differs in having a coarser ornament, no dorsal fold, a weakly impressed muscle field, and a short ventral interarea.

Glyptorthinae sp.
Pl. 4:12

Remarks. – This single, fragmented brachial valve is tentatively placed in the Glyptorthinae on the basis of the impression of the ribs, which are strong and somewhat uneven. The unevenness may be caused by frills.

Family Plectorthidae Schuchert & Le Vene, 1929

Subfamily Plectorthinae Schuchert & Le Vene, 1929

Genus *Plectorthis* Hall & Clarke, 1892

Type species. – By original designation, *Orthis plicatella*, Hall, 1847, p. 122, from the upper Ordovician, Ohio, USA.

?*Plectorthis* sp.
Pl. 4:13, 15–23

Material. – PMO 128.099, 128.100, 128.102, 129.134, 128.149, 128.151, 128.205, 128.210: internal moulds of 4 brachial and 4 pedicle valves from the Myren Member of the Solvik Formation, Vakås, Konglungø, and Spirodden, Asker.

Description. – Exterior: The subquadrate to semi-circular valve outline is $3/5$–$4/5$ as long as wide, with maximum width about half way between the hinge line and the mid-length of the valve. The slightly convex brachial valve has the greatest convexity in the posterior $1/3$ of the valve and a gentle sulcus. The pedicle valve is moderately convex with a fold. The hinge line is gently curved and about $4/5$ as long as the maximum width. The cardinal angles are rounded and the lateral margins are relatively straight, while the anterior margin is evenly curved. The commissure is crenulated and gently uniplicate. The apsacline ventral interarea is low and plane with a narrow (45°), open delthyrium. The dorsal delthyrium is very low and plane with an open notothyrium.

The ribbing is costellate, with 2–3 costella per millimeter at the 5 mm growth stage. Each costellae is angular, with deep interspaces wider than the costellae. Very fine fila occur in the interspaces. Sparse branching is observed. The only exterior expression is a 5 mm long fragment from the umbonal part of a brachial valve, it is therefore not known if the ribbing style agrees with the costate style of *Plectorthis*.

Interior of pedicle valve: The delthyrium chamber is narrow with a small pedicle callist. Small, blunt, and triangular teeth are supported basally by strong but short dental plates. The dental plates continue anteriorly as thick, well-developed dental ridges that initially run parallel and anteriorly curve to define an elongate, cordate to subquadrate muscle field, occupying $2/5$ of the total valve length. The muscle field is deeply impressed with long, slightly crescent-like diductor scars not enclosing the adductor scars that are situated on a long narrow ridge. The mantle canal system is not known.

Interior of brachial valve: The brachiophores are short, thick blades supported by strong, laterally splaying brachiophore plates with converging bases. The shallow sockets have a thin socket pad. The cardinal process is ponderous and bilobate with a crenulated surface. Its highest point is at the mid-length, and it reaches almost to the anterior end of the brachiophore bases. The cardinalia occupy about $1/5$–$1/6$ of the total valve length. A well-impressed, quadripartite muscle field occupies $1/3$ of the valve width and $2/5$ of the total valve length. The anterior scars are subcircular and larger than the triangular posterior scars. The scars are separated by a broad, shallow median ridge. The mantle system is saccate.

Remarks. – The material is sparse (Table 16) and fragmentary, so there has been no attempt to assign it to a species. The assignment to *Plectorthis* is questioned, because the present material lacks concave fulcral plates that seem to be present in typical *Plectorthis*; also lacking is adequate information on the external ribbing. A ribbing style with secondary ribbing occurring away from the umbo is, however, found in several species of *Plectorthis*, e.g., *P. compacta*, Cooper, 1956, *P. australis* Cooper, 1956, and *P. transversa* Cooper, 1956. This may be the youngest *Plectorthis* found so far, occurring in the transitional beds between the Ordovician and Silurian.

Table 16. ?Plectorthis sp., measurements of ventral and dorsal valves.

Ventral valves	s.l.	m.w.	h.w.
PMO 128.149 internal mould	9.00	–	–
PMO 128.151 internal mould	7.00	10.00	6.00
PMO 128.205 internal mould	10.00	12.00	8.00
PMO 128.210 internal mould	10.00	–	–
Dorsal valves	s.l.	m.w.	h.w.
PMO 128.099 internal mould	7.00	11.00	8.00
PMO 128.100 internal mould	7.00	10.00	8.00
PMO 128.102 internal mould	5.00	–	–
PMO 128.134 internal mould	–	14.00	10.00

Plectorthid indet.
Pl. 5:1–8, 11

Material. – PMO 109.762, 128.103, 128.127, 128.130, 128.131, 128.135, 128.136, 128.206, 139.137: internal moulds of 4 brachial and 5 pedicle valves from the lower parts of the Myren Member, Solvik Formation, at Spirodden, Konglungø, Nesøya, and Vakås.

Remarks. – Four fragmentary brachial valves have cardinalia of plectorthid character, e.g., thick, blade-like brachiophores with strong convergent supporting plates and a pronounced bilobed cardinal process. They all seem to have very fine ribbing, and they vary from strongly to moderately convex. Probably the four valves all belong to the same species. The strongly convex nature suggests *Schizophorella*, but a pronounced brachial fold seems to be lacking.

There are a number of unidentified, fine-ribbed pedicle valves from the same beds near the Ordovician–Silurian boundary. They clearly belong to two or three species. There are no criteria whereby complementary brachial valves can be identified. (This material is figured together, even though all of it may not belong to the Plectorthida.)

Subfamily Platystrophiinae Schuchert & Le Vene, 1929

Genus *Platystrophia* King, 1850

Type species. – By original designation, *Terebratulites biforatus* von Schlotheim, 1820, from the Ordovician of the Baltic area.

Platystrophia brachynota (Hall, 1843)

Pl. 5:9–10, 12–20

Synonyms. – □1843 *Delthyris brachynota* – Hall, p. 71, Fig. 6. □1871 *Orthis biforata* Schlotheim – Davidson, p. 268 (*pars*), Pl. 38:20, 21, 24. □1919 *Platystrophia brachynota* (Hall) – McEwan, p. 408, Pl. 42:25–28. □1950 *Platystrophia brachynota* (Hall) – Whittard & Barker, p. 558, Pl. 5:6–14. □1978 *Platystrophia brachynota* (Hall, 1843) – Cocks, p. 56. □1982 *Platystrophia brachynota* (Hall, 1843) – Thomsen & Baarli, Pl. 1:6.

Type occurrence. – From Llandovery beds (late Aeronian–early Telychian), Reynold's Basin, near Niagara, New York, USA.

Material. – PMO 52.595, 105.237, 108.283, 111.686, 128.226, 128227, 128.228 (two specimens), 128.230, 139.138: internal moulds of 5 brachial and 3 pedicle valves and 2 complete valves from the Solvik Formation of the Padda Member, on the west and south coast of Malmøya and the Leangen Member, at Skytterveien, Asker and Kjørbo, Sandvika.

Description. – Exterior: The shell is large, dorsibiconvex, with the pedicle valve $2/3$ as deep as the brachial valve. The outline is transversely subrectangular with a long straight hinge line. The length-to-width ratio varies between $1/2$ and $7/10$. The hinge line varies between $3/5$ and $9/10$ of the maximum width, which occurs in the posterior half of the valve close to the hinge line. There may be an allometric ontogenetic change, with the hinge line becoming longer relative to the maximum width as the shells get larger. The cardinal angles are rounded while the lateral margins are straight to gently curved. The commissure is coarsely crenulated, and the anterior commissure is broadly uniplicate.

The ventral interarea is apsacline, relatively low, and straight, with an open interarea and delthyrial angles of 60–70°. The ventral beak is small and incurved, while the dorsal beak is pronounced, blunt, and incurved up to the hinge line. The dorsal interarea is slightly anacline and low with an open notothyrium.

The ornament is coarsely costellate. The pedicle valve bears a well-defined, deep, angular, and longitudinal median sulcus that originates at the umbo and increases in depth and width anteriorly at about a 45° angle until at the anterior margin it occupies about $1/4$ of the maximum shell width. In the sulcus there are 4 ribs, which occur as 2 ribs near the umbo with the other two ribs originating by division on the

slopes of the median ribs. Each flank of the valve exhibits 7–8 ribs. The brachial valves bear a well defined angular fold with 4–5 ribs. The width or wavelength of the ribs varies, being coarsest medially and becoming finer laterally.

Interior of pedicle valve: The delthyrial chamber is relatively deep with near vertical walls. The teeth are large and blunt, fused laterally with the wall of the delthyrial chamber. The crural fossettes are oblique and shallow on the inner face of the teeth. There is a small, shallow denticular cavity between each tooth and the hinge line. The dental plates are short and descend rapidly to the floor, where they continue in rounded dental ridges that border a raised muscle-bearing platform. The platform is oval to subrectangular, occupying $1/2$–$1/3$ of the total valve length. A weakly impressed muscle field has long narrow diductor scars, each of which is slightly narrower than the long, rectangular adductor scars. The muscle scars may be concentrically striated.

Interior of brachial valve: The notothyrial platform is flanked by thick, short, and divergent brachiophores that are triangular in cross-section. Their median sides are concave, while on the lateral side they define the dental sockets that have small fulcral plates. The cardinal process is a thin, triangular ridge continuing the length of the notothyrial platform. It is flanked on both sides by variably developed notothyrial ridges. The notothyrial platform descends steeply anteriorly into a broad, median ridge. The muscle scars are deeply impressed. The posterior scars are situated lateral of the anterior scars and are oblong with an median depression. They are separated by a broad median ridge, which rapidly dies out and merges into the fold. The anterior muscle scars are less well-defined, probably lobate. Each scar is bisected by a well defined, thin, narrow ridge originating near their posterior ends and continuing along the entire length of the scar.

Remarks. – This material is considered conspecific with *P. brachynota*, as described by Whittard & Barker (1950) from the Bog Quartzite at Shelve, Salop. Table 17 shows, however, that the average size is somewhat larger and the valves are relatively wider than in the material from Salop.

Table 17. Platystrophia brachynota (Hall, 1843), measurements.

Entire valves	s.l.	m.w.	h.w.	m.d.
PMO 52.595	23.00	–	24.00	15.00
PMO 139.138	22.00	35.00	30.00	16.00
Ventral valves	s.l.	m.w.	h.w.	m.d.
PMO 105.237 internal mould	16.00	26.00	16.00	–
PMO 128.228 internal mould	17.00	32.00	25.00	–
Dorsal valves	s.l.	m.w.	h.w.	m.d.
PMO 108.283 internal mould	11.00	21.00	12.00	–
PMO 111.686 internal mould	21.00	32.00	25.00	–
PMO 128.226 internal mould	22.00	–	–	–
PMO 128.230 internal mould	19.00	–	–	–

Family Skenidiidae Kozłowski, 1929

Genus *Skenidioides* Schuchert & Cooper, 1931

Type species. – *Skenidioides billingsi* Schuchert & Cooper, 1931, p. 243; from the Black River Group (Ordovicium, Wilderness Stage) of Paquette Rapids, Ottawa River, Ontario, Canada. By original designation of Schuchert & Cooper (1931, p. 243).

Skenidioides worsleyi sp. nov.

Pl. 6:17–28

Synonymy. □1982 *Skenidioides* sp. – Thomsen & Baarli, Pl. 1:5. □1987 *Skenidioides* sp. – Baarli, Fig. 5j.

Holotype. – PMO 105.892. – Internal mould of a brachial valve from Leangen Member, Solvik Formation, Skytterveien, Asker. Figured by Thomsen & Baarli (1982, Pl. 1:5) as *Skenidioides* sp.

Derivation of name. – Named in honor of Dr. D. Worsley.

Diagnosis. – A *Skenidioides* species with even, fine, and multicostellate ribbing (4–5 ribs per millimeter at 5 mm from umbo) and low and rounded costae. It lacks pronounced ventral fold and has a rounded outline and stout cardinalia. The spondylium simplex is supported by a well-developed septum.

Material. – PMO 105.190, 105.191, 105.192 (two specimens), 105.193, 105.194, 105.892, 109.748, 128.161, 128.163, 128.171, 128.173, 128.174, 128.196, 128.250, 128.317, 136.110, 136.011: internal moulds of 12 pedicle and 5 brachial valves from the Solvik Formation, Myren Member at Spirodden, the Leangen Member at Skytterveien and Leangbukta, Asker, and the Padda Member at Malmøya.

Description. – Exterior: The small, ventribiconvex to near planoconvex shell is $^3/_5$–$^3/_4$ as long as wide, with a moderately high pedicle valve. The outline is semi-circular to subpentagonal. The brachial valve bears a broad, shallow sulcus, seen as a deflection in the posterior part of the shell. In a few valves a deeper, more confined sulcus originates close to the hinge line but becomes shallower towards the anterior margin. The hinge line is straight, $^4/_5$ as long as the maximum width, which occurs between mid-length and the posterior end. The cardinal angle is broadly and evenly rounded, continuing into gently curved lateral margins. The anterior margin is slightly rounded, with the commissure gently and broadly sulcate. The ventral interarea is gently concave and apsacline with open, narrowly triangular delthyrium. The dorsal interarea is very low and anacline, with a open notothyrium medially occupied by the cardinal process.

The radial ornament is finely multicostellate, with 4–5 ribs per millimeter at 5 mm from umbo. The costellae are low, rounded, and equally spaced. A few concentric growth lines are visible.

Interior of pedicle valve: The teeth are relatively small, obtuse, and flatly triangular in cross-section. They are unsupported and attached to the hinge line at the side of the delthyrial opening. The lateral walls of the delthyrium are united to form a spondylium simplex that has a slightly concave floor. The spondylium occupies $^3/_8$ of the total valve length and may extend past the delthyrium. It is supported by a low ridge of varying length arising from the floor of the valve. The ridge is generally well developed, and the spondylium is most often unsupported anteriorly. The vascula media extend as paired trunks anterodistally from the posterior side of the supporting ridge of the spondylium.

Interior of brachial valve: The brachiophores are short, rod-like and greatly divergent (80–90°). They are supported laterally by small fulcral plates. The sockets are shallow, very small, and elevated high above the floor. The brachiophores are supported by stout, slightly concave brachiophore bases, which converge anteromedially onto a longitudinal septum to form a cruralium. The median septum is rounded and highest posteriorly, near the cardinalia. It extends past the muscle field and ends close to the margin of the shell as a low ridge. The cardinal process is a continuation of the median septum and is a simple, fairly thick ridge with its highest point midway along its length. The cardinalia occupy $^1/_4$ of the total valve length.

The muscle scars are well impressed, the oblong anterior scars are the larger, with the elongate posterior scars posterolaterally. The muscle scars occupy about $^1/_2$ of the total valve length and $^1/_3$ of the total valve width.

Remarks. – Tables 18 and 19 show that *S. worsleyi* is a relatively large species of *Skenidioides*. The limited material gives no indications of allometry in relationship of sagittal length and maximum width.

This material clearly falls within the range of variation of the *S. lewisii* group of *Skenidioides*, and it is very close to *S. lewisii* in ribbing. It differs from the Wenlock type material of *S. lewisii* in having a lower pedicle interarea and general valve height, a less pronounced sulcus and no fold, more rounded outline, and stronger impressed muscle scars. *S. hymiri* sp. nov. is also close but for the much coarser ribs that bifurcate more readily than the fine ribs of *S. worsleyi*. *S. worsleyi* has, however, also stouter cardinalia, a longer median ridge, and a much better developed supporting ridge for the spondylium than *S. hymiri*.

Another similar species is the lower Wenlock *S. acutum* (Lindström 1861) from Gotland. *S. acutum* is in need of revision, but according to Whittard and Barker (1950), it has high, pronounced primary ribs and weaker secondary ribs which would set it apart from the new species.

Table 18. Skenidioides worsleyi sp. nov., measurements and statistics for the ventral valves.

	s.l.	m.w.	h.w.
Mean	6.50	9.33	8.25
Std. dev.	1.20	1.87	1.28
Count	8	9	8
Minimum	5.00	8.00	7.00
Maximum	9.00	13.00	10.00
Variance–	1.43	1.86	1.00
covariance		3.50	2.18
matrix			1.64
PMO 105.190 (parat.) int. mould	5.00	8.00	8.00

Table 19. Skenidioides worsleyi sp. nov., measurements and statistics for the dorsal valve.

	s.l.	m.w.	h.w.
PMO 105.892 (holot.) int.mould	4.00	6.00	5.00
Mean	4.25	6.50	5.00
Std. dev.	0.50	0.58	0.00
Count	4	4	2
Minimum	4.00	6.00	5.00
Maximum	5.00	7.00	5.00

The bulk of material containing S. worsleyi originates from the Aeronian Leangen and Padda members of the Solvik Formation. However, one sample of Rhuddanian age from the middle of the Myren Member in Asker has a few seemingly fine-ribbed valves of Skenidioides (PMO 128.171, 173–174) within a population of S. hymiri. Internal characters are much like those of S. worsleyi. However, the one brachial valve present shows better developed and deeper sockets than S. worsleyi; also the anterior muscle scars are bisected by slightly oblique ridges originating near the median septum and oriented anterolaterally (see Pl. 6:23, 27). Tentatively, they are included in S. worsleyi until a larger collection may be obtained.

Skenidioides scoliodus Temple, 1968

Pl. 6:1–6

Synonymy. – □1968 Skenidioides scoliodus sp.nov. – Temple, p. 28, Pl. 5:1–27. □1977 Skenidioides cf. scoliodus Temple 1968 – Havlíček, pp. 101–102, Pl. 33:20. □1986 Skenidioides sp. – Baarli & Harper, Pl. 3a, b. □1990 Skenidioides cf. scoliodus Temple, 1968 – Havlíček & Štorch, p. 49, Pl. 6:11.

Holotype. – By original designation; SM A52188; a brachial valve from Hirnantian mudstones and limestones above Keisley Limestone, near Keisley, Cumbria, Grid ref. NY 714 238.

Material. – PMO 107.609, 107.610, 128.166, 128.179, 128.222–128.225, 136.012, 136.013: internal moulds of 6 pedicle and 4 brachial valves from the lowermost parts of the Myren Member, Solvik Formation, at Vakås, Konglungø, and Spirodden, Asker.

Description. – Exterior. The shells are small with a maximum width of 3.5–7 mm, ventribiconvex, and subpentagonal in outline. The pedicle valve is relatively high, with a pronounced fold and flanks that flatten out laterally from the fold. The brachial valve is moderately convex and about $3/4$ as long as wide. It bears a broad, well-developed sulcus. The hinge line is slightly curved and equal or near in length to the maximum width situated between the hinge line and mid-length of the valve. The cardinal angles are obtusely rounded, while the margins are evenly rounded. The commissure is crenulated and broadly uniplicate. The ventral beak is erect, whereas the ventral interarea is very high, plane, apsacline, $1/2$ as high as wide, and much higher than the low, anacline dorsal interarea. The delthyrium and notothyrium are open, and the delthyrial angle is about 50–60°. The notothyrium is very wide and occupied medially by the cardinal process.

The radial ornament is costate with 15–21 costae along the margins of the valve. External impressions of the valve are not known, but internal impressions of ribbing indicates that the axial rib in the pedicle valve is accentuated and broader than the other ribs. The ribs seem to be rounded, with few secondary branches. In most valves the ribbing is only impressed around the margins.

Internal of pedicle valve: The teeth are moderately small, attached to the anterior margin of the delthyrium, and unsupported basally. The lateral walls of the delthyrium are united medially across the delthyrial cavity to form a spondylium simplex. The spondylium is sessile posteriorly, but free and elevated slightly above the floor anteriorly. The length of the spondylium relative to the delthyrium walls is at least $2/3$, although the exact length is unknown. The spondylium is usually unsupported. The interior of the valve is partially impressed by ribs.

Interior of brachial valve: The brachiophores are short, triangular in cross-section, and diverging at 120°. They are supported laterally by well developed fulcral plates that merge with the hinge line and form bases for relatively shallow, unsupported sockets dipping just barely below the plane of the hinge line. The brachiophores are supported by ventrally concave brachiophore bases that converge anteromedially onto the base of the median septum to form a cruralium. The bases slope down and are without support anteriorly before they rise again to form a moderately high and thin cardinal process that extends along the length of the notothyrial cavity. The median septum is relatively thick and not so high as those of the two other species in the Solvik Formation. It occupies about $4/5$ of total valve length and ascends just anterior of the muscle scars in most specimens. The muscle scars are moderately well impressed and consist of a quadripartite impression with the posterior pair being the smallest. They occupy $7/10$ of the total length and $1/3–2/5$ of the total width.

Table 20. Skenidioides scoliodus (Temple, 1968), measurements and statistics. Four observations were used in the covariance matrix for the pedicle valves, while two cases were omitted due to missing values.

Pedicle valves	s.l.	m.w.	h.w.
Mean	3.4	5.3	5.4
Std. dev.	0.89	1.5	1.7
Count	5	6	5
Minimum	2.00	4.00	4.00
Maximum	4.00	8.00	8.00
Variance–	0.92	1.50	1.50
covariance		3.67	3.67
matrix			3.67

Dorsal valves	s.l.	m.w.	h.w.
PMO 107.610 internal mould	3.50	5.00	4.00
PMO 128.222 internal mould	3.50	5.00	5.00
PMO 128.224 internal mould	4.50	7.00	

Remarks. – The described material is very similar to the type material described by Temple (1968) from Keisley, northern England. Table 20 shows that it is a small species, though larger than the English material. The larger size is the only significant difference and may account for the general lack of impression of ribs on the internal brachial moulds. The Keisley material often displays impressions of ribs, but the larger valves tend to lack it.

Skenidioides hymiri sp. nov.

Pl. 6:7–14

Synonymy. – □1963 *Skenidioides lewisii* (Davidson) – Rubel, 128–129, Pl. 3:1–10. □1982 *Skenidioides woodlandiensis* (Davidson, 1883) – Thomsen & Baarli, Pl. 1:4.

Holotype. – PMO 103.490. A brachial valve from the lower parts of Leangen Member, Solvik Formation, Skytterveien, Asker. Previously figured by Thomsen & Baarli (1982, Pl. 1:4).

Derivation of name. – The species is named after a figure (jotn) in Norse mythology.

Material. – PMO 103.490, 103.491, 128.165, 128.167–128.170, 128.172, 128.175–128.178, 128.267: internal moulds of 8 pedicle and 5 brachial valves from the middle and upper parts of the Myren Member, Solvik Formation, at Spirodden, Asker, and from the basal part of Leangen Member at Vettrebukta. The species also occurs rarely in the basal Spirodden Member at Spirodden.

Diagnosis. – A coarse-ribbed *Skenidioides* (2–3 ribs per millimeter at valve margin) with high, rounded, triangular ribs that often bifurcate near the anterior margin. The sulcus and fold are weakly developed, while the interarea is very high and the spondylium long and unsupported. The cru-

ralium is composed of relatively straight, antero-medially directed brachiophore-supporting plates, fusing with the median ridge.

Description. – Exterior: Ventribiconvex to planoconvex shell, medium sized for the species. The pedicle valve is relatively high and evenly convex. The brachial valve is transversely semi-circular and about $\frac{1}{2}$–$\frac{4}{5}$ (most commonly $\frac{3}{5}$) as long as wide. It may bear a broad, shallow sulcus arising in the anterior parts. The hinge line is straight to slightly curved and equal in length to maximum width. The cardinal angles are obtusely rounded to angular, while the margins are evenly rounded. The commissure is crenulated and gently sulcate.

The ventral beak is erect, while the ventral interarea is plane, apsacline, $\frac{3}{10}$ as high as wide and much higher than the low anacline dorsal interarea. The delthyrium and notothyrium are open, and the delthyrial angle is about 30°. The notothyrium is occupied medially by the cardinal process.

The radial ornament is costate, with an average of 2–3 costae per millimeter and 16–27 costae along the margin of the valves. The ribs are rounded triangular and high, with well developed growth imbrications and few secondary branches.

Internal of pedicle valve: The teeth are very small to near non-existent. They are attached to the anterior margin of the delthyrium and unsupported basally. The lateral walls of the delthyrium are united medially across the delthyrial cavity to form a spondylium simplex with evenly rounded sides and floor. The spondylium occupies the same length as the delthyrium walls and is usually unsupported. One specimen shows rounded semi-circular impressions on the bottom of the valve on either side of the spondylium (Pl. 6:9), similar to those described by Whittard & Barker (1950). Divergent vascula media flank the lateral, median side of these impressions. They do not extend from the impressions as described by Wittard & Barker (1950), rather from the area at the side of the spondylium and immediately posterior to the impressions. From this specimen, it appears as if the mantle system is lemniscate, and the semi-circular impressions are the areas between branched vascula media and weaker vascula genitalia. The interior of the valve is commonly impressed by ribs.

Interior of brachial valve: The brachiophores are short, triangular in cross-section, and diverge at 60°. They are supported laterally by well-developed fulcral plates that merge with the hinge line and form bases for relatively well developed sockets above the valve floor. The brachiophores are supported by brachiophore bases that converge antero-medially onto the base of the median septum to form a cruralium. The bases are not concave, as they are in *S. scoliodus*. The median septum is thin and high and ascends abruptly just past the anterior parts of the muscle scars. It occupies about $\frac{3}{5}$ of the total valve length. Posteriorly, the septum fuses with a fairly high and very thin cardinal process that runs the length of the notothyrial cavity. The muscle scars are lightly impressed and often obscured by overprinted ribs. They consist of a quadripartite impression with the posterior pair being smaller than the more oblong posterior scars. They occupy $\frac{2}{5}$ of the total length and $\frac{1}{4}$ of the total width.

Remarks. – There is a large variation in shape within this limited material (Tables 21, 22). However, no patterns of allometry can be detected.

The material appears similar to that described by Rubel (1963) as *S. lewisii* (Davidson, 1848) from the lower Llandovery of Estonia. This species, however, was rejected from *S. lewisii* by Bassett (1972) in his redescription of the type material. Here I tentatively include it in the new species *S. hymiri.*

The Norwegian specimens seem to display a greater variation in width versus length, but they overlap with the more rounded Estonian specimens. There is also a possibility that *Skenidioides hymiri* is synynomous with the poorly known *Skenidioides woodlandiensis* (Davidson, 1883). However, the present material is different from the *Skenidioides* sp. [*?woodlandiensis* Davidson, 1883] described by Temple (1970). Cocks (1978) synynomized Temple's material with *S. woodlandiensis.* The differences are that *S.* sp. [*?woodlandiensis*] has an angular fold, higher frequency of ribs, and a more transverse outline than the Norwegian and the Estonian material. Internally the spondylium in the British material is shorter and more frequently supported by a ridge.

Table 21. Skenidioides hymiri sp. nov., measurements and statistics of the ventral valve.

	s.l.	m.w.	h.w.
Mean	4.88	6.86	6.80
Std. dev.	1.31	0.38	0.45
Count	4	7	5
Minimum	3.50	6.00	6.00
Maximum	6.00	7.00	7.00
PMO 128.169 (parat.) int. mould	6.00	7.00	7.00
PMO 128.177 (parat.) int. mould	4.00	7.00	7.00

Table 22. Skenidioides hymiri sp. nov., measurements and statistics for the dorsal valve.

	s.l.	m.w.	h.w.
Mean	4.12	6.50	6.25
Std. dev.	1.65	2.38	2.06
Count	4	4	4
Minimum	2.50	4.00	4.00
Maximum	6.00	9.00	8.00
Variance–	2.73	3.92	3.29
covariance		5.67	4.83
matrix			4.25
PMO 103.490 (holot.) int. mould	6.00	9.00	8.00
PMO 128.165 (parat.) int. mould	3.00	5.00	5.00

Skenidioides sp.
Pl. 6:15, 16

Remarks. – The material consists of a single brachial valve from the upper parts of the Llandovery at Malmøya. It is distinct in its double ridges where the brachial process should have been.

Order Strophomenida Öpik, 1934

Suborder Strophomenidina Öpik, 1934

Superfamily Plectambonitacea Jones, 1928

Family Leptestiidae Öpik, 1933

Genus *Leangella* (*Leangella*) Öpik, 1933

Type species. – *Plectambonites scissa* (Salter) var. *triangularis* Holtedahl, 1916, p. 84, Pl. 15:5, 6, from the Silurian (Llandovery, Leangen Member, Solvik Formation) railroad section near Asker Farm, Asker, Norway. By original designation of Öpik (1933, p. 42).

Leangella scissa (Davidson, 1871) *triangularis* (Holtedahl, 1916)
Pl. 7:1–11

Type specimen. – Lectotype selected here as the original mold of a pedicle valve shown in Holtedahl's Pl. 15:6.

Synonymy. – □1916 *Plectambonites scissa* Salter var. *triangularis* Holtedahl – Holtedahl, pp. 84–85, Pl. 15:5, 6. □1933 *Leangella triangularis* (Holtedahl) – Öpik, pp. 42–48, Pl. 8:6–8, Pl. 9:1–4. □1970 *Leangella triangularis* (Holtedahl) – Cocks, p. 158. □1982 *Leangella scissa* (Davidson) – Thomsen & Baarli, Pl. 1:17. □1986 *Leangella scissa* (Davidson) – Baarli & Harper, Pl. 2c. □1987 *Leangella scissa* (Davidson) – Baarli, Fig. 5f.

Material. – PMO 105.226 (21 specimens), 128.264–128.266, 128.268, 128.269, 128.271, 128.272, 128.273 (2 specimens), 128.274 (9 specimens), 128.275 (2 specimens), 128.276 (4 specimens), 128.277 (7 specimens), 128.278 (5 specimens), 128. 279 (2 specimens), 128.281, 128.282 (2 specimens; this slab also contains 26 external moulds of pedicle valves, which are hard to measure because the margin of the shells are seldom preserved), 128.285 (3 specimens), 128.344, 128.355, 128.370, 128.371, 130.906 (7 specimens), 130.907 (5 specimens), 130.908 (6 specimens), 130.909, 130.910: internal moulds of 45 pedicle and 27 brachial valves and external moulds of 9 brachial and 5 pedicle valves from the Myren Member at Konglungø and the Leangen member at

Skytterveien and Bleikerveien, Asker. This species is ubiquitous in the Solvik Formation, occurring at all levels and at all localities where the Solvik Formation is exposed.

Description. – Exterior: Strongly to moderately concavo-convex with the curvature of the pedicle valve greater than that of the brachial valve. The pedicle valve, especially in large specimens, is strongest curved posteriorly. The outline is semi-circular through subtriangular to pear-shaped, with maximum width at the hinge line. The cardinal angles are acutely rounded and often moderately alate. The ratio of width to length varies from 2:1 in small valves to 1.2:1 in large specimens (see Fig. 5). The lateral and anterior margins vary according to outline. The commissure is smooth and rectimarginate. The umbo is weak, while the ventral interarea is narrow and triangular, apsa- to anacline. The dorsal interarea is straight, strongly hypercline and about $\frac{1}{3}$ as high as the ventral interarea. The delthyrial angles are 60–70°, with a mesothyridid foramen. The pseudodeltidium is gently convex and extends along the delthidial margin as slender plates. The delthyrium is occupied by the trilobed cardinal process which is flanked by chilidial plates.

The ornament is parvicostellate with low, fine primary ribs varying in number from 5 to 18 in a population of 21 pedicle valves showing an average of 9.9 ribs. The dorsal valves show a variation between 5 and 13 ribs, with an average of 9 ribs in a population of six. All the valves came from the same bulk sample at 16 m above the base of the Leangen Member, Solvik Formation, Skytterveien, Asker. The secondary ribs are extremely fine and number from 5 to 10 between each primary rib. In one specimen rugae are developed near the umbo.

Interior of pedicle valve: The teeth are strong and widely separated. They are supported by slender dental plates which diverge laterally onto the valve floor. The delthyrium is filled apically by a massive apical thickening. High ridges run anteriorly from the dental plates and turn sharply to circumscribe the muscle scars. They may sometimes curve posteriorly again into an invagination defining a broad heart-shaped muscle field. The muscle field is strongly impressed, bilobed and situated on the apical thickening. The thickening occupies $\frac{1}{4}$–$\frac{1}{5}$ of the total length and $\frac{1}{3}$ of the total width. The adductor scars are small, oval and confined to the posterior part of the apical thickening. These scars are longitudinally bisected by a slender myophore that stops posteriorly at the pedicle callist. The diductor scars are relatively large, rounded and situated anterolaterally to the adductor scars that they confine. The diductor scars may show concentric growth-marks or they may be longitudinally striated, probably by the posterior part of the mantle-canal system. Some specimens (Pl. 7:9), especially the larger ones, have developed a subperipheral ridge close to the lateral and anterior margins. Clearly impressed vascula media originate at the points of the heart-shaped apical thickening on each side of the invagination. The vascula media branch shortly

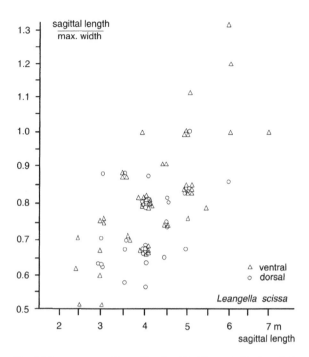

Fig. 5. Measurements of *Leangella scissa* (Davidson, 1871) *triangularis* (Holtedahl, 1916), showing how the ratio of the sagittal length to maximum width increases with sagittal length.

anteriorly and send one branch towards the median ridge, while the other branch curves down, defining an oblong field between them.

Interior of brachial valve: The cardinal process is strong and trilobed, with the median lobe dominant, filling the notothyrial opening. The socket ridges are continuous with the lateral part of the cardinal process. These ridges are strong, low, and flared laterally, nearly parallel to the hinge line. Dental sockets are fairly deep. The bema is well developed and bilobed by a broad axial indentation. It projects freely anteriorly, and its anterior base makes a posteriorly directed W, although the lower parts of the W may be variously drawn out and accentuated. The bema occupies $\frac{2}{5}$–$\frac{1}{2}$ of the maximum width and $\frac{3}{5}$–$\frac{3}{4}$ of the maximum length. It is larger relative to the total valve length in small specimens. The lateral sides of the bema are undulated, and medially there is a double, raised ridge. Many of the valves show radial striations on the anterior parts of the bema. The vascula media are either parallel or curved pairs of trunks (first converging then diverging) projecting anteriorly from the tips of the bema lobes.

A platform extends from the hinge line and parallel to the margins. However, some show an invagination anteromedially away from the margin, making a pear-shaped platform. The platform is variably developed from a high or even undercut rim to a slight slope towards the margin. In the few available valves from the lower parts of the Solvik Formation, the positive rim is changed to a rounded trough. Typical for these populations are stunted or very small forms, 2–4 mm wide. The main part of the studied material comes from the Leangen Member (6c, where the type *L. triangularis* was collected). These valves are typically large for the species (5–8 mm wide) and occur abundantly.

Remarks. – Tables 23 and 24 show that *Lengella scissa triangularis* is a small species and that the covariance between sagittal length and maximum width is very low, in spite of an adequate number of specimens. The scattergram (Fig. 5) reveals a great variation in shape at each growth-level, which explains some of the low covariance. However, there is also a clear increase in the ratio between sagittal length and maximum width with increasing sagittal length, e.g., the valves become relatively longer as they mature.

Holtedahl (1916) described populations as *Plectambonites scissa* Salter var. *triangularis* from the same horizon in Asker where I collected the majority of specimens. Öpik (1933) elevated this subspecies to species rank. As noted by Holtedahl (1916) and Öpik (1933), the Norwegian populations typically show a higher number of ribs than the British *L. scissa*. The numerical range of ribs varies between 5 and 18, with an average of near 10 on the pedicle valve and 9 in the brachial valve. Cocks (1970) noted a variation between 4 and 14 ribs in the British material, while Temple (1970) stated on basis of a large population from Mathrafal, Montgomeryshire, that *L. scissa* had up to 9 ribs in the brachial valve and only 5 ribs in the pedicle valve. Clearly Holtedahl's and Öpik's observations are confirmed. The other main reason

for separating the Norwegian material as a different species, besides the ribbing, was that *L. triangularis* had an invagination of the platform creating a pear shape, which Öpik (1933) claimed *L. scissa* lacked. The abundant Norwegian material at hand shows all degrees of invagination, though most specimens lack it. Cocks (1970), Cocks & Rong (1989), and Temple (1970) all figured specimens of *L. scissa* with well developed pear-shaped platforms. Thus, the variation within populations and overlap in features are so great that there is no reason to distinguish the Norwegian material from the British material at the species level. The differences in rib frequency, however, support Holtedahl's (1916) subspecies.

Family Xenambonitidae Cooper, 1956

Subfamily Aegiromeninae Havlíček, 1961

Genus *Aegiria* (*Aegiria*) Öpik, 1933

Type species. – *Aegiria norvegica* Öpik, 1933, p. 55, Pls. 10:1–5; 11:3–5, from the Solvik Formation (Lower Llandovery), Leangen, Asker, Norway.

Aegiria norvegica Öpik, 1933
Pl. 7:12–23

Synonymy. – □1933 *Aegiria norvegica* n.sp. – Öpik, p. 55, Pls. 10:1–5; 11:3–5. □1982 *Aegiria norvegica* Öpik – Thomsen & Baarli, Pl. 1:18. □1987 *Aegiria norvegica* Öpik – Baarli, Fig. 5e. □1987 *Aegiria norvegica* Öpik – Cocks & Rong, p. 122.

Type material. – Same as type species.

Material. – PMO 103.489, 103.505, 103.574, 128.338, 128.340, 128.342, 128.343, 128.346 – 128.348, 128.350, 128.352 (2 specimens), 128.253, 128.354, 128.356 – 128.360, 128.362 – 128.364, 128.372 (3 specimens), 136.014, 136.015: internal moulds of 10 brachial and 13 pedicle valves and 5 external moulds of valves from the middle and upper parts of the Myren Member, Spirodden, and the basal parts of the Leangen Member, Skytterveien. *A. norvegica* is also found in the Spirodden Member at Spirodden and as a rare taxon in the Myren Member at Malmøya.

Description. – Exterior: The shells are small and slightly concavo-convex to near planoconvex, with the pedicle valve being the deeper. The outline is transversely semi-circular to semi-elliptical with the length half of the width. There is a slight fold in the pedicle valve and a slight sulcus in the brachial valve, both of which originate near the umbo. The hinge line is straight, and the cardinal angles are obtuse but may be slightly alate in large specimens. The maximum width is at the hinge line or slightly anterior of it. The lateral margins are evenly rounded and the anterior margin is

Table 23. Leangella scissa (Davidson, 1871) *triangularis* (Holtedahl, 1916), measurements and statistics for the ventral valves.

	s.l.	m.w.	h.w.
Mean	4.19	5.16	5.16
Std. dev.	1.02	0.85	0.85
Count	44	43	43
Minimum	2.50	3.50	3.50
Maximum	7.00	7.00	7.00
Variance–	1.04	0.49	0.49
covariance		0.72	0.72
matrix			0.72

Table 24. Leangella scissa (Davidson, 1871) *triangularis* (Holtedahl, 1916), measurements and statistics of the dorsal valves.

	s.l.	m.w.	h.w.
Mean	4.03	5.44	5.49
Std. dev.	0.55	0.83	0.80
Count	35	36	35
Minimum	3.00	4.00	4.00
Maximum	5.00	7.50	7.50
Variance–	0.31	0.23	0.23
covariance		0.68	0.64
matrix			0.64

slightly rounded to nearly straight. The commissure is crenulated and gently unimarginate.

The umbo is inconspicuous and not incurved. A fairly high, triangular interarea in the pedicle valve is plane to weakly concave and apsacline. The brachial valve has a lower, hypercline interarea. A small pseudodelthidium is present close to the apex. The open chilidium has walls that diverge at 90° and is filled with the three-lobed cardinal process.

The ornament is finely multicostellate to fascicostellate, with 6–7 rounded costae per millimeter at the anterior commissure. The costae increase in number by lateral branching.

Interior of pedicle valve: The hinge line is smooth, except for small, bluntly rounded teeth situated close to the delthyrium. The teeth are supported by short and widely splayed dental plates continuous with short, rounded muscle-bounding ridges that curve medially in large specimens to define the muscle field laterally and anteriorly. The delthyrial chamber is shallow. A small, well-developed pedicle callist is present. A flat ridge extends anteriorly from it, bisecting the small, oval adductor scars. The diductor scars are variably impressed but clearly visible in large specimens. They are diverging and pentagonal in outline, situated lateral and anterior to the adductor scars. The external ornament is impressed on the moulds, obscuring mantle canal systems. Rounded, elongate papillae are present in the anterior half of the valves.

Interior of brachial valve: The cardinal process is small and trifid. The sockets are shallow and bordered by robust socket plates that are widely splayed, almost parallel to the hinge line, and continuing far past the sockets. At the base of the cardinal process is a small rounded depression. Anterior of the hollow is a strong median ridge dividing the muscle scars and continuing slightly past the bema. The bema is semioval, well developed, and undercut in all specimens. The rim of the bema varies from smooth to extended in several small, radial ridges, which are seen especially in large specimens. The muscle field is radially striated.

Remarks. – Tables 25 and 26 show that *Aegiria norvegica* is a small species with relatively small variation in shape through development. There is some variation in shape within each growth level, as reflected in the slightly low covariance, especially for the ventral valves. Some of this variation may, however, be explained by errors in measurements, since this is such a small species.

Temple (1987) regarded *A. norvegica* as a synonym of *A. garthensis* (Jones, 1928), while Öpik (1933) distinguished them on the basis of frequency and style of ribbing and degree of undercutting of the bema. As pointed out by Temple (1987), the ribbing is similar, although *A. norvegica* may have ribs with a stronger tendency to cluster in bunches approaching a fascicostellate style. Temple (1987) also regarded the strong undercutting of the bema as a mere function of larger size in the Norwegian material. However, this feature is seen in all brachial valves and, in fact, the average size of the British (sagittal length 3.37 mm, maximum width

Table 25. Aegiria norvegica (Öpik, 1933) measurements and statistics for the ventral valves.

	s.l.	m.w.	h.w.
Mean	3.46	6.92	6.83
Std. dev.	0.75	1.73	1.70
Count	13	12	12
Minimum	2.00	3.00	3.00
Maximum	4.00	9.00	9.00
Variance–	0.56	1.18	1.14
covariance		2.99	2.89
matrix			2.88

Table 26. Aegiria norvegica (Öpik, 1933) measurements and statistics for the dorsal valves.

	s.l.	m.w.	h.w.
Mean	4.00	6.57	6.57
Std. dev.	1.00	2.07	2.07
Count	9	7	7
Minimum	3.00	5.00	5.00
Maximum	5.00	10.00	10.00
Variance–	1.00	2.13	2.10
covariance		4.29	4.30
matrix			4.30

7.9 mm for 3 valves) and Norwegian (sagittal length 4.0 mm, maximum width 6.57 mm for 9 and 7 valves, respectively) material are close. These figures, together with similar and corresponding numbers for the ventral valves, may indicate that the British *A. garthensis* is wider relative to length than the Norwegian *A. norvegia*. Following Cocks & Rong (1989), I therefore retain Öpik's species name, although the two species are clearly similar.

Family Sowerbyellidae Öpik, 1930

Subfamily Sowerbyellinae Öpik, 1930

Genus *Eoplectodonta* (*Eoplectodonta*) Kozłowski, 1929

Type species. – *Sowerbyella precursor* Jones, 1928, a junior subjective synonym of *Leptaena duplicata* J. de C. Sowerby, 1839.

Eoplectodonta duplicata (J. de C. Sowerby, 1839)

Pls. 7:25–26, 28–29; 8:1–7, 10, 13, 16

Synonymy. – □1839 *Leptaena duplicata* – J. de C. Sowerby *in* Murchison, p. 636, Pl. 22:2. □1848 *Leptaena transversalis* var. *undulata* – Salter *in* Phillips & Salter, p. 372. □1916 *Plectam-*

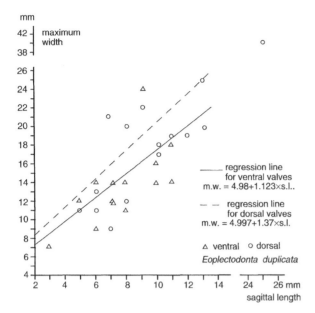

Fig. 6. Scattergram with regression lines for ventral and dorsal valves of *Eoplectodonta duplicata* (J. de C. Sowerby, 1839) plotting sagittal length against maximum width.

bonites transversalis Wahl – Holtedahl (*pars*), pp. 83–84, Pl. 15:2–4 non Pl. 15:1. □1917 *Plectambonites transversalis* (Dalman) – Reed, p. 886, Pl. 15:35, 36. □1917 *Plectambonites transversalis* var. *duvali* (Davidson) – Reed, p. 887, Pl. 15:37–40. □1917 *Plectambonites transversalis* var. nov. *mullochensis* – Reed, p. 887, Pls. 15:41, 42; 16:1, 2. □1917 *Plectambonites transversalis* var. nov. *tricostata* – Reed, p. 889, Pl. 16:8–13. □1928 *Sowerbyella duplicata* (J. de C. Sowerby) – Jones, p. 432, Pls. 22:6–13; 23:1, 2. □1928 *Sowerbyella precursor* sp. nov. – Jones, p. 437, Pl. 23:3–5. □1928 *Sowerbyella mullochensis* (Reed) – Jones, p. 439, Pl. 23:6–9. □1928 *Sowerbyella superstes* sp. nov. – Jones, p. 441, Pl. 23:10–12. □1928 *Sowerbyella undulata* (Salter) – Jones, p. 452, Pl. 24:3–6. □1928 *Sowerbyella undulata* var. *maccoyi* var. nov. – Jones, p. 457, Pl. 24:7. □1928 *Sowerbyella undulata* var. *tricostata* (Reed) – Jones, p. 458, Pl. 24:8, 9. □1970 *Eoplectodonta duplicata* (J. de C. Sowerby) – Cocks, p. 169, Pls. 5:1–12; 6:1–13; 7:1–11; 8:1–11. □1970 *Ygerodiscus undulata* (Salter) – Cocks, p. 185, Pl. 14:3–12, Pl. 15:1–12. □1970 *Eoplectodonta duplicata* (J. de C. Sowerby) *undulata* (Salter) – Temple, p. 40, Pls. 10:1–13; 11:1–16. □1978 *Eoplectodonta duplicata* (J. de C. Sowerby) – Cocks, p. 100. □1978 *Ygerodiscus undulata* (Salter 1848) – Cocks, p. 103. □1981 *Eoplectodonta undulata* (Salter) – Williams & Wright, p. 23, Fig. 7. □1982 *Eoplectodonta duplicata* (J. de C. Sowerby, 1839) – Thomsen & Baarli, Pl. 1:8. □1983 *Eoplectodonta* cf. *duplicata* (J. de C. Sowerby) – Lockley, p. 94, Fig. 2. □1984 *Eoplectodonta duplicata* (J. de C. Sowerby) – Temple *in* Cocks *et.al.*, pp. 150, 152–154, Figs. 20–22. □1986 *Eoplectodonta duplicata* (J. de C. Sowerby) – Baarli & Harper, Pl. 2g, h, i, and j. □1987 *Eoplectodonta duplicata* (J. de C. Sowerby) – Baarli, Fig. 5h. □1987 *Eoplectodonta dupli-*

cata (J. de C. Sowerby) – Temple, pp. 60–70, Pl. 6:1–17. □1989 *Eoplectodonta duplicata* (J. de C. Sowerby, 1839) – Cocks & Rong, p. 135, Figs. 128, 129.

Type material. – Lectotype selected by Jones (1928, p. 432); GMS Geological Society Collection 6874, internal mould of pedicle valve, the original of J. de C. Sowerby *in* Murchison 1839, Pl. 22:2, from Rhuddanian beds at Cefn Rhyddan, Llandovery, Dyfed, Grid ref. SN 762 325.

Material. – PMO 103.479, 103.530, 103.532, 105.881, 128.164, 128.268, 128.270, 128.288 (2 specimens), 128.289 – 128.293, 128.294 (2 specimens), 128.295 – 128.297, 128.298 (2 specimens), 128.299 – 128.307, 128.308 (3 specimens), 128.309 – 128.312, 128.313 (2 specimens), 128.314 – 128.316, 128.318 – 128.320, 128.351, 128.369, 130.897: internal moulds of 22 pedicle and 9 brachial valves and exterior moulds of 17 valves from the Myren Member at Spirodden, Konglungø, and Vakås, the Leangen Member, at Skytterveien, and the Padda Member, at Malmøya, all the Solvik Formation.

Description. – Exterior: The valves are moderately to strongly concavo-convex, with highest convexity posterior on the pedicle valve. The ventral beak is swollen and often somewhat curved over the hinge line. This is most pronounced in larger valves. The dorsal beak is inconspicuous. The outline is transverse, with maximum length relative to width varying between $\frac{2}{5}$ and $\frac{4}{5}$ (commonly $\frac{3}{5}$) as long as wide (see Fig. 6). The average ratio of the total population is 0.56. Lateral and anterior margins are generally evenly curved, but a few specimens have alae. The anterior margin is rectimarginate, although a few specimens display undulations of the shell resulting in gentle folds and sulci. The ventral interarea is apsacline to anacline and about equal in length to the dorsal interarea. The delthyrium is open, with a delthyrial angle of about 120°. The larges group of the specimens (40%) have an open delthyrium, some (30%) have small deltidial plates, while the rest have only a slight thickening of the delthyrial wall (Pl. 8:1, 4, 7). The dorsal interarea is hypercline, with a notothyrium usually having two well-developed chilidial plates, although some specimens (Pl. 8:1, 7) have an entire chilidium. A three-lobed cardinal process with a strongly developed median lobe may be seen in the notothyrial opening.

The ornament is parvicostellate, with a variable number (12–30) of primary costae. Also the number of costellae is variable, from 7 to 20, in the few valves where costellae are observed. A few valves are uniformly costellate, since the primary costae are lacking. Others have a prominent central costa.

Interior of pedicle valve: The hinge line is denticulated for $\frac{1}{2}$–$\frac{3}{4}$ of its length, with rounded denticles gradually decreasing in size laterally. The dental plates are variably developed, small to inconspicuous thin ridges that delimit the muscle field and fade within a short distance anteriorly. The muscle field is variably outlined and variably impressed. The pedicle callist is well developed. A stout, short median ridge divides the adductor scars in most specimens and sometimes ends in bifurcation. The adductor scars are well impressed, small, and drop-shaped to rectangular. They are flanked posteriorly by flaring diductor scars. The shape is oval and often two-lobed to multilobed anteriorly. The mantle system is often strongly impressed, pinnate with a pair of thick, variably diverging vascula media originating at the anterolateral margins of the diductor scars. The areas outside the muscle scars bear coarse, elongately rounded papillae.

Interior of brachial valve: The hinge line bears complimentary fosettes to the denticles in the pedicle valve. The cardinal process is well developed, trilobed with a blunt, rounded median lobe and two lower side lobes, and it juts posteriorly of the hinge line. The side lobes are fused with clavicular plates laterally. The clavicular plates are variably developed; either they are thin plates flaring laterally at 120° angle to each other or they are larger and curve medially into swellings or pads where they merge with the median septum. In the latter case, they border a flat hollow on the anterior side of the cardinal process. The median septum is usually well developed, relatively broad, and reaching a variable distance anteriorly. The inner side septa are well developed and strong, continuing forward for ⅔ of the total length of the valve. The outer-side septa are variable but always present. They are strongest and highest medially. The bema is also variable, from almost invisible to well marked and broadest at the mid-length of the shell. The sides of the bema are thin and running nearly parallel before they curve back posteriorly into two lobes. The mantle canal system is generally not impressed.

Remarks. – Tables 27 and 28 show that *Eoplectodonta duplicata* is a medium-sized species with a wide variation in measurements. The variation is especially well seen in the data for the ventral valves (Table 27, Fig. 6) where the covariance between sagittal length and maximum width is low, and a relatively wide scatter of shape for each growth stage is evident. However, the regression lines also indicate a certain degree of allometric growth where the width increases slightly slower than the sagittal length. The valves become relatively longer when they mature.

That *E. duplicata* is a very variable species is also demonstrated by Cocks (1970) and, especially, Temple (1987). In the Norwegian material there are all gradations between *E. undulata*-like specimens and specimens without undulation, which supports Temple's (1987) conclusion that *E. undulata* Jones, 1928, is a synonym of *E. duplicata*. Cocks (1970) stated that there are no delthidial plates in *E. duplicata*. Temple (1970) noted thickening of the walls at the side of the delthyrium but did not report deltidial plates. The Norwegian material shows all types of variation from very well developed deltidial plates to no traces at all; also the chilidial plates vary from relatively small to total cover. There were no stratigraphical or topical patterns in these details. The variable feature of deltidial plates is not enough

Table 27. Eoplectodonta duplicata (J. de C. Sowerby, 1839), measurements and statistics for the ventral valves.

	s.l.	m.w.	h.w.
Mean	8.18	14.37	14.06
Std. dev.	2.88	4.45	4.36
Count	17	19	18
Minimum	3.00	7.00	7.00
Maximum	14.00	24.00	24.00
Variance–	8.28	6.12	6.12
covariance		19.80	19.00
matrix			19.00

Table 28. Eoplectodonta duplicata (J. de C. Sowerby, 1839), measurements and statistics for the dorsal valve.

	s.l.	m.w.	h.w.
Mean	9.67	18.78	18.50
Std. dev.	4.50	6.96	7.59
Count	18	18	16
Minimum	5.00	9.00	8.00
Maximum	25.00	40.00	40.00
Variance–	20.24	31.09	34.14
covariance		48.42	55.50
matrix		57.60	

to separate the Norwegian material from *E. duplicata*. For further discussion, see under *E. transversalis jongensis* subsp. nov.

Eoplectodonta transversalis (Wahlenberg 1818) *jongensis* subsp. nov.

Pl. 8:8–9, 11–12, 14–15, 17–18, Pl. 9:1, 4

Type specimen. – The holotype of *E. transversalis jongensis* is an internal mould of brachial valve PMO 135.963 from the uppermost parts of the Rytteråker Formation, Kampebråten, Sandvika.

Diagnosis. – A small subspecies with valves displaying low frequency of costa without prominent growth lines, a relatively high length/width relationship, and a deep, inrolled valve with a small interarea.

Derivation of name. – From Jong, the area around Kampebråten where the subspecies occurs abundantly.

Synonymy. – □1916 *Plectambonites transversalis* Wahl – Holtedahl (*pars*), pp. 83–84, Pl. 15:1, non 2–4. □1982 *Eoplectodonta penkillensis* (Reed 1917) – Cocks & Baarli, Pl. 1:1–9.

Material. – PMO 111.665 (2 specimens), 111.700 (2 specimens), 130.935, 130.936 (3 specimens), 130.937, 135.909, 135.911, 135.912, 135.925, 135.948, 135.954, 135.956 (3 specimens), 135.958, 135.959 (2 specimens), 135.960, 135.963, 135.964, 135.986: internal moulds of 5 pedicle and

10 brachial valves and external moulds of 6 brachial and 1 pedicle valve, from the uppermost parts of Rytteråker Formation at Kampebråten, Sandvika and Christian Skredsviks vei, Bærum. The species also occurs in the lower parts of the Vik Formation at the same localities and in the Vik Formation on Malmøya and Malmøykalven.

Description. – Exterior: The valves are small and strongly convex, although the convexity varies. The ventral beak protrudes in front of and overhangs the hinge line. The dorsal beak may also overhang slightly and shows the strongest convexity adjacent to the hinge line. The outline is very variable, with length relative to width being between $\frac{1}{3}$ and $\frac{5}{6}$. The average ratio for the whole population is 0.61. The outline is semicircular to subquadrate. Alae may rarely be present. The lateral and anterior margins are generally evenly curved, with the anterior margin rectimarginate. Only one specimen (Pl. 9:1) shows small undulations of the shell.

The ventral interarea is apsacline and relative short. The delthyrium is open, and the angle is about 120°. Poor preservation prevents knowledge of delthyrial plates. The dorsal interarea is anacline to hypercline, often with a partially closed chilidium where only the upper part of the median lobe of the cardinal process is visible. Some chilidium plates may possibly be complete.

The ornament is generally parvicostellate, with 8–11 primary costa and 6–13 costellae between the primary costae. These numbers are obtained from a small sample and may not be representative of the total variation. A few valves are close to uniformly costellate, since the primary costa are weakly developed. Small rugae may be present near the lateral parts of the hinge line.

Interior of pedicle valve: The hinge line is denticulated with tiny denticles. The dental plates are small to nearly nonexistent, very thin, and diverging strongly on each side of the muscle field where they very soon die out. The muscle field is variably impressed. The adductor scars are very small, dropshaped, and situated on either side of a well developed median ridge that originates anterior of a small pedicle callist. The median ridge is of variable length and often bifurcates anteriorly. The adductor scars are flanked anterolaterally by relatively small and variably impressed diductor scars. Anteriorly, the diductor scars continue into a pair of diverging and strongly impressed vascula media. The trunks branch at about $\frac{2}{3}$ length of the valve. The area outside the muscle scar is pitted by rounded papillae.

Interior of brachial valve: The hinge line bears complementary fosettes to the denticles in the pedicle valve. The cardinalia are well developed, trilobed with clavicular plates diverging at 110°. The clavicular plates border a deep, hollow pit on the anterior side of the cardinal process. The pit is smaller than that of *E. duplicata.* The median septum is well developed and often reaches $\frac{2}{3}$ of the valve length anteriorly. The side septa are well developed and often surpass the length of the median septum. The outer septa are short but

Table 29. Eoplectodonta transversalis (Wahlenberg, 1818) *jongensis* subsp. nov, measurements and statistics for the ventral valve.

	s.l.	m.w.	h.w.
Mean	4.75	7.33	7.33
Std. dev.	1.54	1.03	1.03
Count	6	6	6
Minimum	3.00	6.00	6.00
Maximum	7.00	9.00	9.00
Variance–	2.38	0.80	0.80
covariance		1.07	1.07
matrix			1.07

Table 30. Eoplectodonta transversalis (Wahlenberg, 1818) *jongensis* subsp. nov., measurements and statistics for the doral valves.

	s.l.	m.w.	h.w.
Mean	3.93	6.83	6.69
Std. dev.	0.65	0.79	0.75
Count	14	15	13
Minimum	2.50	5.00	5.00
Maximum	5.00	8.00	8.00
Variance–	0.42	0.06	0.06
covariance		0.40	0.33
matrix			0.33

always present. The bema is variably developed and often highest laterally to the outer septa. The mantle canal system is not impressed.

Remarks. – Tables 29 and 30 show that this is a small species with relatively little variation in size range. There is, however, great morphological variation within each size level. Consequently, the covariance is extremely low, expressing the large plasticity in shape. There may also be some elements of allometric growth, as in *E. duplicata,* but nearly all the length measurements cluster around 4 mm (Fig. 7). Thus the regression lines are very sensitive to the shape of the few valves with outlying plots. This has to be held in mind when interpreting growth variation. The width shows very little increase with increase in sagittal length.

The present material falls within the morphological range of *E. transversalis* although showing minor differences that justify the erection of a subspecies. Cocks (1970) treated both *E. transversalis transversalis* from Gotland and *E. transversalis penkillensis* from Great Britain. *E. transversalis jongensis* is distinguished from both in a lower frequency of costa. From the former it further differs in having a smaller interarea and lack of prominent growth lines. From the latter it differs in having a higher length/width ratio and deeper and stronger enrolled valves.

Both *E. transversalis* and *E. duplicata* are highly variable in morphology, and the differences between the two species are not always clear. However, Fig. 8 shows that in the Norwe-

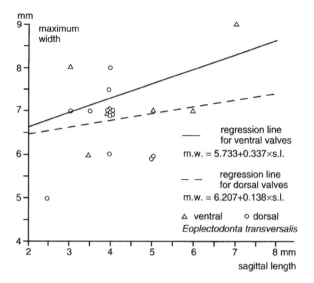

Fig. 7. Scattergram with regression lines for ventral and dorsal valves of *Eoplectodonta transversalis* (Wahlenberg, 1818) *jongensis* subsp. nov., plotting sagittal length against maximum width.

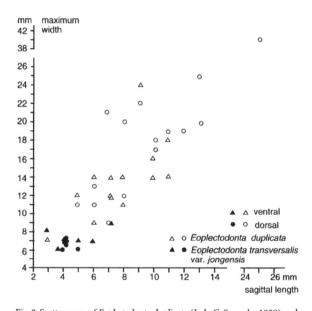

Fig. 8. Scattergram of *Eoplectodonta duplicata* (J. de C. Sowerby 1839) and *Eoplectodonta transversalis* (Wahlenberg 1818) *jongensis* subsp. nov., showing plots of sagittal length against maximum width for the two populations.

gian material there is a very clear difference in size without overlap in mature specimens. There are other more gradational differences, but they all show overlaps. Unfortunately, the only definite difference stated by Cocks (1970), the presence or absence of delthyrial plates, is not valid in the Norwegian material since specimens of both species may have these, not only *E. transversalis*.

?*Sowerbyella* sp.
Pl. 7:24, 27

Remarks. – Only one small brachial valve from the lowermost Myren Member at Ostøya was found. The apparent lack of denticles and the two prominent side septa led to placement with *Sowerbyella*, but the material is not satisfactory for a definite determination.

Superfamily Strophomenoidea Öpik, 1934

Family Strophomenidae King, 1846

Subfamily Furcitellinae Williams, 1965

Genus *Katastrophomena* Cocks, 1968

Type species. – By original designation, *Strophomena antiquata* var. *woodlandensis* Reed, 1917. Lectotype selected by Cocks (1968, p. 295); B 54490; a pedicle valve, the original of Reed, 1917, Pl. 18:1, from the Woodland Formation (Rhuddanian), Woodland Point, Girvan, Strathclyde. Grid ref. NX 168 952.

Katastrophomena woodlandensis (Reed, 1917)
Pl. 9:3, 5–18

Synonymy. – □1871 *Strophomena antiquata* (J. de C. Sowerby) – Davidson, p. 299 (*pars*), Pl. 44:21, 22, *non* Pl. 44:2–13. □1883 *Strophomena antiquata* (J. de C. Sowerby) – Davidson, p. 193, Pl. 15:12–14. □1916 *Strophomena? costatula* Hall and Clarke – Holtedahl, pp. 62–63, Pl. 13:6, 7. □1916 *Strophomena* sp. – Holtedahl, p. 63, Pl. 13:8. □1917 *Strophomena antiquata* (J. de C. Sowerby) var. *woodlandensis* – Reed, p. 902, Pls. 18:20, 21; 19:1–5. □1949 *Strophomena scotica* – Bancroft, p. 13, Pl. 1:6, 7. □1951 *Strophomena scotica* (Bancroft MS) sp. nov. – Williams, p. 116, Pl. 7:1–3, Text-figs. 19a–b. □1951 *Strophomena* aff. *woodlandensis* Reed – Williams, p. 118, Pl. 7:4. □1968 *Katastrophomena scotica* (Bancroft) – Cocks, p. 296, Pl. 3:3–9. □1968 *Katastrophomena woodlandensis* (Reed) – Temple, p. 44, Pl. 8:16, 17. □1970 *Katastrophomena woodlandensis* (Reed, 1917) – Temple, p. 44, Pl. 8:16–17. □1978 *Katastrophomena scotica* (Bancroft, 1949) – Cocks, p. 109. □1978 *Katastrophomena woodlandensis* (Reed, 1917) – Cocks, p. 109. □1982 *Katastrophomena scotica* (Bancroft, 1949) – Thomsen & Baarli, Pl. 1:19. □1987 *Katastrophomena woodlandensis* (Reed, 1917) *scotica* (Bancroft, 1947) – Temple, pp. 72–75, Pl. 7:1–14. □1989 *Katastrophomena scotica* (Bancroft, 1949) – Kul'kov & Severgina, p. 131, Pl. 27:9, 10.

Type specimen. – See under type for genus.

Material. – PMO 40.097, 51.976, 53.607 (2 specimens), 128.238, 128.252, 130.353, 130.355, 130.357 – 130.363, 130.918: internal moulds of 6 pedicle and 5 brachial valves and external moulds of 4 valves from the basal Myren Member, Solvik Formation, at Semsvannet, Nesøya, Konglungø, Spirodden, and Ostøya, and from the basal Leangen Member at Vettrebukta, Asker.

Description. – Exterior: The outline is subcircular to subtriangular, with maximum length varying from $3/5$ to $4/5$ of maximum width, which occurs at the straight hinge line. Small rounded alae may be present. The valves are variably resupinate from convexoplane to sharply geniculate. The brachial valve is variable convex. The larger valves are relatively flat in the posterior half and then geniculate sharply to produce the strongest convexity anteriorly. Lateral margins vary from nearly straight to evenly rounded with the anterior margin. The anterior margin may, however, be drawn out in a lip with a ventral sulcus and corresponding dorsal fold. The umbo is low, and the ventral beak is erect, with a sealed foramen.

The ventral interarea is plane, triangular, and apsacline, three to four times as long as the thin, anacline dorsal interarea. The delthyrium is sealed by a large, convex pseudodeltidium. The notothyrium is closed by a small, convex chilidium.

The radial ornament is coarsely and unevenly costate with unevenly high costa. The growth lines are often prominent, producing uneven nodes on the ribs. The rib density 10 mm from umbo is 15–18 ribs per 10 mm. The ribs originate mainly by intercalations, but branching is also seen.

Interior of pedicle valve: The delthyrial chamber is relatively shallow, with a gently concave floor. The teeth are well developed and bluntly rounded, with dorsal faces lying in the plane of the interarea. The dental plates are short and continue as high, curved muscle-bounding ridges that are variably undercut. The ridges curve towards the median and are often drawn out in variable long subparallel ridges on each side of a well-developed median ridge. The median ridge is often broad and flanked by the thin, blade-like adductor scars, which often reach well anterior of the diductor scars. The diductor scars are broadly triangular, with concentric ridges. A pedicle callist is present.

Interior of brachial valve: The cardinalia are raised on a short, high notothyrial platform. The two cardinal-process lobes are short, oval, stout and dorsoposteriorly directed. They are placed closely together with a thin groove between them. The socket plates are widely splayed and mostly straight, although the lateral edges may be curved. The muscle field, which is contained posteriorly by the socket plates, is variably impressed and subcircular to triangular in outline. The adductor scars are elongate drop-shaped to lanceolate, and situated on each side of a variable median ridge. The median ridge varies from a long, strong ridge

Table 31. *Katastrophomena woodlandensis* (Reed, 1917), measurements of valves.

Ventral valves	s.l.	m.w.	h.w.
PMO 51.976 external mould	17.00	20.00	20.00
PMO 53.607 external mould	15.00	23.00	23.00
PMO 53.607 internal mould	14.00	21.00	21.00
PMO 128.238 internal mould	–	20.00	–
PMO 130.918 internal mould	–	20.00	20.00

Brachial valves	s.l.	m.w.	h.w.
PMO 128.252 internal mould	29.00	34.00	34.00
PMO 130.360 internal mould	–	32.00	32.00

reaching beyond the muscle field, through a short, slightly bifurcated ridge to a faint, short depression right anterior of the notothyrial platform. The diductor scars are variable but larger than and bounding the adductor scars laterally. The trans-muscle septa are always present. The muscle-bounding ridges vary but may be strongly developed.

Remarks. – There is not enough material for a statistical treatment, but Table 31 shows the measurements. The strong ribbing and the development of strong growth lines and nodes in the Norwegian material are reminiscent of the Wenlock species *K. antiquata* (J. de C. Sowerby, 1839). However, it lacks the fine costella seen in that species. Bassett (1974), in discussing *K. antiquata*, commented on the great variation in ribbing styles and speculated that the ribbing was environmentally determined. Similarly, on the grounds of ribbing density Temple (1987) concluded that the two species *K. scotia* and *K. woodlandensis* are partly separated as two end members of a continuum. He consequently changed the status of *K. scotia* to a subspecies of *K. woodlandensis*. The difference in outline of the ventral muscle-bounding ridges he attributed to different modes of preservation; this interpretation is not supported here. The Norwegian material has the coarse ribbing of *K. woodlandensis woodlandensis* but most often the same kind of muscle-bounding ridges as *K. woodlandensis scotica*. Consequently, a new subspecies could be erected. However, with all the variation observed in the Norwegian material, I think that naming a new subspecies would be of little value.

Katastrophomena penkillensis (Reed, 1917)
Pl. 10:1–19

Lectotype. – Selected by Cocks (1968, p. 298); B 73013; internal mould of a brachial valve; the original of Reed, 1917, Pl. 18:11, from the Wood Burn Formation (Telychian) of Bargany Pond Burn, near Girvan, Strathclyde. Grid ref. NX 250986.

Synonymy. – □1871 *Strophomena antiquata* (J. de C. Sowerby) – Davidson, p. 299 (*pars*), Pl. 44:5, *non* Pl. 44:2–4, 6–13, 21, 22. □1916 *Strophomena antiquata* (J. de C. Sowerby) – Holtedahl, pp. 58–60 (*pars*), Pl. 10:6–8, *non* Pl. 10:9, 10. □1917 *Strophonella penkillensis* – Reed, p. 900, Pl. 18:11–14. □1968 *Katastrophomena penkillensis* (Reed) – Cocks, p. 296, Pl. 3:3–9. □1982 *Katastrophomena* cf. *penkillensis* (Reed, 1917) – Thomsen & Baarli, Pl. 1:11, 12.

Material. – PMO 103.502, 103.544, 103.637, 105.227, 105.870, 105.875, 105.879, 105.882, 105.885, 128.240, 128.241, 128.243, 128.249, 128.323–128.325, 128.327–128.332, 128.334–128.336, 128.372 (2 specimens), 128.382–128.384, 128.399, 130.309–130.316, 130.318–130.320, 130.341, 130.382, 130.911, 130.912: internal moulds of 12 pedicle and 21 brachial valves and 13 external moulds from the Leangen Member, Solvik Formation of Skytterveien, Jongsåsveien, and Leangbukta. *K. penkillensis* is also found in the Bruflat Formation at Purkøya in the Ringerike District.

Description. – Exterior: The outline is elongately oblong to subtriangular, ¾ to equally as wide as long. Larger specimens tend to be longer than wide. The maximum width occurs most often at the hinge line, but larger specimens might have maximum width farther anterior towards mid-length. The pedicle valve is resupinate, nearly flat to gently convex near umbo. The brachial valve varies from flat to convex, with change to ventrally directed growth at about 1 cm growth stage. The hinge line is straight and sometimes extended into small alae. The lateral and anterior margins are evenly curved, with the anterior margin rectimarginate. In larger specimens the anterior margin can be gently uniplicate. The umbones are low, with the ventral beak erect. The ventral interarea is straight, relatively low, plane, and apsacline. The delthyrium is wide, with angles of 100–120°. The dorsal interarea is about as long as the ventral interarea. The pseudodeltidium is relatively large, while the chilidium is small; both are convex.

The radial ornament is parvicostellate with 12–14 accentuated costae per 10 mm at 5 mm from the umbo. The number of finer ribs between the costae varies from 5 to 11. Rugae are present, often strongly developed, in small specimens near the posterior margin. Some specimens show rugae interrupted by major costae at an acute angle producing an ornament somewhat like *Gunnarella* Spjeldnæs, 1957 (Pl. 10:4). In larger valves the rugae are not so strongly impressed and the zig-zag pattern is missing, probably because of increased valve thickness.

Interior of pedicle valve: The delthyrial chamber is relatively shallow, with strong, short dental plates supporting knob-like, triangular teeth with dorsal faces in the plane of the hinge line. The denticles observed by Temple (1987) on *K. woodlandensis* are not developed in this material, in spite of excellent preservation. The dental plates continue as

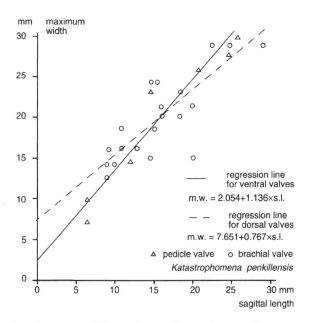

Fig. 9. Scattergram with regression lines for ventral and dorsal valves of *Katastrophomena penkillensis* (Reed, 1917), plotting sagittal length against maximum width.

strong muscle-bounding ridges that sometimes are undercut and splay out laterally. The ridges curve evenly towards the median ridge without meeting it. In a few valves the muscle-bounding ridges curve to continue subparallel with the median ridge. The muscle field is subcircular to subquadrate or diamond-shaped. A well-developed rounded median ridge extends from the pedicle callist to the muscle-bounding ridges or slightly anterior of them. Thin, deeply impressed, lanceolate adductor scars flank the ridge and sometimes flare out slightly anteriorly. They surpass the large triangular diductor scars that may be either concentrically or longitudinally striated. The ribbing and rugae are variably impressed, but they are stronger impressed on smaller and presumably thinner shells.

Interior of brachial valve: The cardinalia are raised on a variably developed notothyrial platform. The platform is well developed in large specimens but nearly lacking in small specimens. The cardinal-process lobes are stout, tubular, and directed ventriposteriorly. There is a thin, fine cleft between them. The socket plates are strong, straight, and widely diverging (120°), extending anterolaterally beyond the generally faintly impressed muscle scars. A median ridge is variably developed, most often present in large individuals, and lacking in juvenile valves. One valve shows a bifurcate median ridge. The trans-muscle septa are most often well developed. Only two valves have deeply impressed muscle fields. One shows muscle fields bounded by sinuous muscle-bounding ridges, and the muscle field occupies $\frac{1}{3}$ of total width and $\frac{2}{3}$ of the length. In another valve the muscle field occupies $\frac{1}{3}$ of both length and width.

Remarks. – This form was originally ascribed by Baarli (1987) to two separate taxa, *K.* cf. *penkillensis* and *Pentlandina* sp. The material ascribed to *K.* cf. *penkillensis* consisted of large specimens from several localities, while the material of *Pentlandina* sp. was a relatively large collection of small specimens from one locality where *K.* cf. *penkillensis* also occurred. The small specimens were mostly transverse, flat, or only gently geniculate with strongly impressed ribbing, sharp rugae, and a faintly developed notothyrial platform. Few pedicle valves were present, but one of them gave an impression of a ventral sulcus, thus all the small specimens were placed in *Pentlandina*. Closer investigation showed that the small specimens were juvenile forms of *K. penkillensis*, and Fig. 9 shows that there is no break in plots of shape between the two populations. Tables 32 and 33 reveal that the covariance between sagittal length and maximum width is somewhat low, especially for the dorsal valves, where the number of specimens is adequate. Inspection of Fig. 9 shows that with a few exceptions there is relatively little variation within each growth stage. The regression lines corroborate the impression of broad juvenile valves that gradually develop into more equal length and width towards maturity. The change in growth producing geniculation occurs at about the 1 cm

growth stage (e.g., Pl. 10:7) and later growth occurs more anteriorly, thus producing an entirely different outline in mature specimens than in young ones. The thickening of the shell with age obliterates the rugae.

The British material investigated by Cocks (1968) is scarce and fairly poorly preserved. It naturally shows less variation but agrees in inner structures with the Norwegian material. The Norwegian and British material agree very well in terms of their unique ornament. There is thus no reason to keep it under open nomenclature. The British material, however, is limited in age to Late Llandovery, while the Norwegian material is from strata correlated with the lower half of Aeronian as well as the upper half of the Telychian. In Norway there is some stratigraphic overlap with the dominantly Ruddanian *K. woodlandensis* and the largely younger *K. penkillensis*. They both occur in the coeval lowermost parts of the Leangen Member of earliest Aeronian age, but at different locations. *K. woodlandensis* occurs in a more distal, lime-rich setting than *K. penkillensis*, which typically occurs in argillaceous tempestite environments. A brief investigation of occurrences from Britain shows that the Rhuddanian and Aeronian *K. woodlandensis* is succeeded without overlap by the Telychian *K. penkillensis* and that the relationships to sedimentary setting seems to be the same as in Norway. The distribution may therefore be at least partly environmentally determined, which agrees with Bassett's (1974) assessment of *K. antiquata*.

Table 32. Katastrophomena penkillensis (Reed, 1917), measurements and statistics for the ventral valves.

	s.l.	m.w.	h.w.
Mean	15.14	18.00	18.57
Std. dev.	7.49	8.18	8.66
Count	7	8	7
Minimum	6.00	7.00	7.00
Maximum	26.00	30.00	30.00
Variance–	56.14	70.27	84.80
covariance		66.86	74.95
matrix			74.95

Table 33. Katastrophomena penkillensis (Reed, 1917), measurements and statistics for the dorsal valves.

	s.l.	m.w.	h.w.
Mean	15.48	19.62	18.94
Std. dev.	5.46	5.07	4.14
Count	21	21	18
Minimum	9.00	12.00	12.00
Maximum	29.00	29.00	26.00
Variance–	29.86	23.27	18.44
covariance		25.75	19.21
matrix			17.11

Genus *Dactylogonia* Ulrich & Cooper, 1942

Type species. – Original designation: *Dactylogonia geniculata* Ulrich & Cooper (1942), USNM 108202a p. 623, from the Airline Formation (middle Ordovician) of Tennessee, USA.

Dactylogonia dejecta sp. nov.
Pl. 11:1, 5, 8, 11, 15–16

Synonymy. – □1982 *Cyphomena* sp. – Thomsen & Baarli, Pl. 2:3.

Holotype. – The inner mould of brachial valve PMO 105.231, from 11 m above the base of the Leangen Member, Solvik Formation, Skytterveien, Asker.

Derivation of name. – Latin *dejecta*, cast down, sad.

Material. – PMO 105.231 (2 specimens), 105.232, 128.161, 128.365, 128.386: internal moulds of 3 pedicle and 2 brachial valves and 1 external mould of a brachial valve, all from the Leangen Member, Solvik Formation of Skytterveien, Asker.

Diagnosis. – This is a *Dactylogonia* species with subparallel, anteriorly directed muscle-bounding ridges in the ventral interior valve and smooth teeth braced by tiny accessory plates. The ventral valve shows sharp geniculation, while the dorsal valves are concave. The cardinalia are delicate, and the inner trans-muscle septa are long, low, thin, and subparallel, bordering the median side of the anterior muscle scars. The outer trans-muscle septa are very short and close to the inner septa.

Description. – Exterior: The outline is semi-circular to transversely subquadrate. The valves are concavo-convex, with the brachial valve initially flat to gently concave, showing the strongest concavity in the anterior $\frac{1}{5}$ of the valve. The pedicle valve is gently convex posteriorly, with a sharp geniculation in dorsal direction. The maximum width occurs at the hinge line, which may be produced into small, rounded ears. The lateral and anterior margins are evenly rounded, and the anterior margin is rectimarginate. The umbones are small, with a small open pedicle foramen. The ventral interarea is flat and two to three times as high as the low dorsal interarea. There is a small, convex pseudodeltidium partly closing the delthyrium. The delthyrial angle is about 120°. The chilidium is convex, well developed, and drawn down in a small projection between the cardinal-process lobes.

The ribbing, which is seen only on one valve, is very fine and equally multicostellate, with 5–7 costa per millimeter at 10 mm from umbo. The preservation is not good, but the costae seem to be low, rounded and somewhat irregular in thickness. This irregularity may be due to nodes or fila. Rugae are weakly developed but present all over the disc.

Interior of pedicle valve: The delthyrial chamber is very shallow. The teeth are triangular, smooth, and well developed with posterior faces in the plane of the hinge line. They are braced by tiny accessory plates. The dental plates are short but stout, with bases passing into low, anteriorly directed dental ridges that extend parallel to each other anteriorly. The direction of the ridges may be deflected anteriorly at an angle, so that they converge closer to the median ridge. A thin median ridge is present, reaching half-way between the anterior ends of the muscle-bounding ridges. The muscle scars are faintly impressed, but the diductor scars are probably triangular to trapezoid and do not encircle, the small, thin and lanceolate to triangular adductor scars. There is a well-developed subperipheral flange in the posterior parts of the valve.

Interior of brachial valve: The cardinalia are on a slightly raised and triangular notothyrial platform. The cardinal-process lobes are stout, elongate, and parallel with a groove between them. Their posterior faces are covered by the chilidium. The socket plates are thin, strongly divergent, and slightly curved laterally. The plates are set at an angle of about 30° to the hinge line. The socket plates are not crenulated. They contain the muscle field that is weakly impressed except

Table 34. Dactylogonia dejecta sp. nov., measurements for the valves.

Ventral valves	s.l.	m.w.	h.w.
PMO 105.232 (parat.) int. mould	14.00	21.00	21.00
PMO 128.386 int. mould	17.00	22.00	22.00

Dorsal valves	s.l.	m.w.	h.w.
PMO 105.231 (holot.) int. mould	16.00	24.00	24.00
PMO 105.231 (parat.) ext. mould	20.00	24.00	24.00
PMO 128.161 int. mould	16.00	21.00	21.00

for small, elongate adductor scars that are seen posteriorly on each side of a broad, median ridge. The ridge fades at about $\frac{1}{4}$ of the valve length from the posterior. Two deep and strong, trans-muscle septa extend to about half the valve length. They are slightly sinuous and do not converge anteriorly. The posterior ends are in front of the notothyrial platform, and they border the inside of the muscle scars. A second pair of very short ridges are found lateral to and near the posterior end of the longer pair.

Remarks. – This species is very rare, but good preservation justifies erection of a new species. The measurements for the available material are given in Table 34. Possibly this is the same species as *Biparetis* sp. of Temple (1987). The pedicle muscle-bounding ridges and the notothyrial platform seem different in the two taxa, but again both descriptions are based on only a few specimens. *Biparetis* Amsden, 1974, is found in the Hirnantian Lemon Formation of Missouri while *Dactylogonia*, widespread in North America, is usually of middle Ordovician age. The present material seems to occupy a place between the two morphologies. The main features distinguishing the two genera are the possession of two trans-muscle septa, lack of dental accessory plates, and smooth teeth and sockets in *Dactylogonia* while *Biparetis* has only one pair of trans-muscle septa, accessory plates and crenulated teeth and sockets. *D. dejecta* has two trans-muscle septa, but the outer pair is very short and so close as to nearly merge with the inner pair. It possesses tiny accessory plates, but the teeth and sockets are smooth. Faintly impressed pedicle muscle scars unite *D. dejecta* with *Dactylogonia*, although the nearly quadrate shape of the muscle field in *D. dejecta* separates it from both genera. Amsden (1974, p. 55) noted that *Dactylogonia* lacks rugae in contrast to *Biparetis*. This is contrary to the definition of Ulrich & Cooper (1942). They stated that *Dactylogonia* usually lacks wrinkles but nevertheless describe several species with rugae, for example *D. magnifica* Cooper, 1956, that in external shape and possession of rugae closely resemble the Norwegian material. In spite of the fact that the Norwegian material is removed from *Dactylogonia* both in time and, to a lesser degree, in morphology, I retain the name because of the presence of outer trans-muscle septa.

Family Rafinesquinae Schuchert, 1893

Subfamily Leptaeninae Hall & Clarke, 1894

Genus *Leptaena* Dalman, 1828

Type species. – Designated by King (1846, p. 37): *Leptaena rugosa* Dalman, 1828, p. 106, from the Upper Ordovician of Västergötland, Sweden.

Leptaena haverfordensis Bancroft, 1949
Pl. 11:2–4, 6–7, 9–10, 12–14, 17

Synonymy. – □1949 *Leptaena haverfordensis* sp. nov. – Bancroft, p. 6, Pl. 1:19, 20, 23, 24. □1949 *Leptaena haverfordensis* var. *contracta* – Bancroft, p. 6., Pl. 1:21, 22. □1968 *Leptaena haverfordensis* Bancroft – Cocks, p. 304, Pl. 5:4–15. □1978 *Leptaena haverfordensis* Bancroft, 1949 – Cocks, p. 115. □1982 *Leptaena haverfordensis* Bancroft, 1949 – Thomsen & Baarli, Pl. 2:2–5. □1987 *Leptaena haverfordensis* Bancroft, 1949 – Temple, pp. 75–78, Pl. 8:1–10. □1989 ?*Leptaena* cf. *haverfordensis* Bancroft, 1949 – Kul'kov & Severgina, p. 137, Pl. 28:1.

Type material. – Lectotype designated by Cocks (1968, p. 304). Internal mould of ventral valve, SM A32163, from Gasworks Mudstone (Haverford Mudstone Formation), opposite entrance to Gasworks, Haverfordwest, Dyfed (Bancroft 1949, Pl. 1:20; Cocks 1968, Pl. 5:4, Pl. 8:1).

Material. – PMO 103.478, 103.511, 108.273, 128.231–128.234, 128.237–128.239, 128.242, 128.244, 128.258, 128.261–128.263, 128.286, 128.287, 128.345, 128.351, 128.366, 130.349, 130.914, 130.917: internal moulds of 4 pedicle and 4 brachial valves, external moulds of 5 brachial and 4 pedicle valves, and 3 complete valves, from Solvik Formation, Asker, in the Myren Member at Nesøya, Konglungø, and Spirodden, Asker, and the Leangen Member, Vettrebukta and Skytterveien. *L. haverfordensis* is also found in the Spirodden Member at Spirodden and throughout the Solvik Formation at Sandvika.

Description. – Exterior: The outline of the shell is subcircular to subrectangular and resupinate. Maximum width is at the hinge line, which often continues laterally as pronounced ears on large valves. The angle of geniculation varies between 70° and 110°. Some small valves have trails extended antero-medially into a point, as noted by Temple (1987). In the pedicle valve, the periphery of the disc at its junction with the trail is formed into a variably pronounced rim with a complementary hollow in the dorsal valve. The ventral beak bears a small round foramen, commonly sealed. The umbo is low and inconspicuous, with a ventral beak that is short and erect. The ventral interarea is apsacline, low, triangular and horizontally striated. The delthyrium is wide, with angles of 90–100° and a convex pseudodeltidium. The dorsal interarea

is also low and anacline. The chilidium is large in comparison to the pseudodeltidium, convex, and extending ventrally beyond the hinge line.

The number of rugae on the disc varies between 7 and 12. They are absent from the trail and less pronounced near the umbo. The radial ornament is finely multicostellate, with 3–4 costa per millimeter at the 10 mm growth stage.

Interior of pedicle valve: The delthyrial cavity is shallow and of variable width and length with a pedicle callist. The teeth are fairly small, bluntly triangular with posterior faces flush with the wall of the delthyrium. The dental plates are well developed, but short. They are fused anteriorly with strong muscle-bounding ridges that curve evenly and most often meet medially. Some muscle scars may be slightly flabellate. The muscle scars are deeply impressed, occupying $\frac{1}{6}$–$\frac{1}{3}$ of the total disk width and $\frac{1}{2}$–$\frac{4}{5}$ of the total disc length. The diductor scars are broadly oval to triangular, often enclosing well-impressed oval to lanceolate adductor scars. The adductor scars are situated on a platform divided by a thin median ridge extending just anterior of the muscle scars. The extra muscular area is coarsely tuberculate and often overprinted by rugae.

Interior of brachial valve: The cardinalia are relatively strong. The cardinal-process lobes are strongly elongate, fairly thin, and extending anteriorly at a slight angle. The lobes are widely separated. The notothyrium platform is weakly developed, if present. The socket ridges are very thin and widely divergent. The dental sockets are shallow, small, and subparallel to the hinge line. The muscle scars are well

Table 35. Leptaena haverforensis (Bancroft, 1949), measurements and statistics for the ventral valves.

	s.l.	m.w.	h.w.
Mean	10.78	18.90	18.90
Std. dev.	1.86	4.01	4.01
Count	9	10	10
Minimum	7.00	13.00	13.00
Maximum	13.00	25.00	25.00
Variance–	3.44	5.08	5.08
covariance		16.10	16.10
matrix			16.10

Table 36. Leptaena haverforensis (Bancroft, 1949), measurements and statistics for the dorsal valves.

	s.l.	m.w.	h.w.
Mean	13.83	17.44	17.44
Std. dev.	2.79	2.92	2.92
Count	6	9	9
Minimum	12.00	12.00	12.00
Maximum	19.00	21.00	21.00
Variance–	7.77	3.03	3.03
covariance		8.53	8.53
matrix			8.53

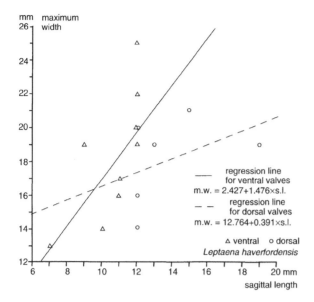

Fig. 10. Scattergram with regression lines for ventral and dorsal valves of *Leptaena haverforensis* Bancroft, 1949, plotting sagittal length against maximum width.

developed. Rounded, broad posterior scars are divided by a low median ridge. The smaller, elongate, and oval anterior scars are divided anteriorly by a thin and sometimes deep myophragm that continues anterior of the muscle scars. The scars occupy $\frac{1}{2}$–$\frac{2}{5}$ of the total disc width and $\frac{1}{4}$ of the disc length.

Remarks. – Tables 35 and 36 give the measurements and statistics for *L. haverfordensis*. It is a medium sized species. The covariance between sagittal length and maximum width is very low. Fig. 10 shows that there is an enormous spread in shape for each growth stage. The regression lines plot very different for dorsal and ventral valves. This may be explained by a predominance of plots within a narrow length interval, with very few outlying plots. The regression lines are therefore strongly influenced by a few plots. These may lead to very different angles of the lines if they are from valves of different shape.

The material agrees well with the British type material of this species. For further comments, look under *L. valentia*.

Leptaena valida Bancroft, 1949

Pl. 12:5–9, 11–12, 14

Synonymy. – □1949 *Leptaena valida* sp. nov. – Bancroft, p. 6, Pl. 1:25. □1949 *Leptaena elongata* sp. nov. – Bancroft, p. 7, Pl. 1:26, 27. □1968 *Leptaena valida* Bancroft – Cocks, p. 305, Pl. 6:1–5. □1978 *Leptaena valida* Bancroft, 1949 – Cocks, p. 117. □1982 *Leptaena valentia* Cocks, 1968 – Thomsen & Baarli, Pl. 2:1.

Type specimen. – Lectotype of *L. valida*, selected by Cocks (1968, p. 305); SM A35690; a pedicle valve, the original of Bancroft 1949, Pl. 1:25, from C1 Beds (Aeronian), River Sefin, SE of Lletyr'hyddod, near Llandovery, Dyfed. Grid ref. SN 742 282.

Material. – PMO 105.195, 128.245, 128.251, 128.257, 130.348, 130.350, 130.351, 130.915, 130.916: internal moulds of 3 pedicle and 1 brachial valve, external moulds of 4 pedicle valves, and 1 conjoined pair of valves, from the top of the Spirodden Member, Solvik Formation, at Spirodden and the base of the Leangen Member, Solvik Formation, at Vettrebukta and Skytterveien, Asker.

Description. – Exterior: The shell is fairly thick with a slightly transverse, subangular outline. The valves are resupinate with long trails. Some valves show an asymmetrical outline of the disc, with the trail better developed and with a different angle to the disc on one side than on the other. The pedicle valve is differentiated from other *Leptaena* species by a relatively pronounced beak that continues in an anteriorly directed thickened section running over the disc, creating a wedge that stands slightly up from the rest of the disc. The exterior of the brachial valve is not known. The disc is about $\frac{3}{4}$ to equally as long as wide. The maximum width of the disc occurs at the hinge line, where small ears may be developed. The angle of geniculation varies from 80° to 120° and is generally in the higher range. There may be a pronounced rim in the pedicle valve at the periphery of the disc, but this is generally not well developed.

The ventral beak bears a small open foramen. The umbo is small, although more prominent than in many species of *Leptaena*. The ventral interarea is flat and apsacline. The nature of the delthyrium is not known. The dorsal interarea is straight, low and anacline. The chilidium is small.

The disc is interrupted by 8–10 well developed and regular rugae. The rugae are absent from the trail and increase in size from the umbo. The radial ornament is finely multicostellate, with a triangular wedge of 5–8 more pronounced and higher costellae at the middle part of the disc, creating a parvicostellate pattern in this section. These stronger costa continue onto the trail. There are about 5 costae per millimeter at the 10 mm growth stage.

Interior of pedicle valve: The delthyrial cavity is broad and shallow, with a well-developed pedicle callist. The dental plates are very short to obsolete and fused basally to very strong dental ridges that initially curve evenly towards the median ridge but subsequently change direction to create two narrow lobes anteriorly in front of the median ridge. The muscle field is deeply impressed, large and cordate to lobate. It occupies $\frac{1}{3}$–$\frac{2}{5}$ of the total valve width and about $\frac{1}{2}$ of the total length measured on the two available shells. The adductor scars are lanceolate and located on either side of a well-developed median ridge. The diductor scars are fan-shaped and radially striated. Faint impressions of parallel vascula

Table 37. Leptaena valida (Bancroft, 1949), measurements and statistics of valves.

	s.l.	m.w.	h.w.
Mean	13.71	16.71	16.71
Std. dev.	4.46	7.97	7.97
Count	7	7	7
Minimum	9.00	10.00	10.00
Maximum	23.00	33.00	33.00
Variance–	19.90	33.74	33.74
covariance		63.57	63.57
matrix			63.57

media may be traced to the anterior edge of the disc. The rugae are visible interiorly.

Interior of brachial valve: This description is based on one single, fragmented shell. The cardinalia are relatively small compared with other *Leptaena*. They are situated on a slightly elevated, broad platform that extends fairly far anteriorly, continuing as a broad but short median ridge. The cardinal-process lobes are low, rounded and divided by a thin, shallow groove. The chilidium is drawn down in a small lobe between the processes. The faces of the cardinal-process lobes are directed posteroventrally, and the lobes are slightly divergent to subparallel. The sockets are shallow and bounded by very thin, low socket ridges, which diverge at 100–120°. The muscle field is not readily observed. Rugae and ribs are overprinted on the inside of the valve.

Remarks. – Table 37 gives the measurements for the valves and statistics for the ventral valves. The material is sparse, but the covariance between sagittal length and maximum width is good.

Leptaena valida was first defined on a few contorted pedicle valves from strata in the lower parts of the Upper Llandovery sequence at Sefin River, Llandovery. The Norwegian material is scarce and from a limited horizon around the base of the Leangen Member in transitional Rhuddanian to Aeronian strata. The combination of a pronounced wedge with parvicostellate ribbing on the median part of the pedicle valve and lobate pedicle muscle-bounding ridges, however, is seen in no other *Leptaena*. The Norwegian material differs from the Welsh specimens in having a less transverse outline, but there is some variation in this feature. The main locality for *L. valida*, Vettrebukta, also yielded a few specimens of *L. haverfordensis*, but *L. valida* dominated. The one brachial valve found at this locality differs from *L. haverfordensis* in the closely adjacent, small, and rounded cardinal-process lobes, compared with the widely separated and very elongate lobes of *L. haverfordensis*. There are thus good reasons to assume that this brachial valve represents *L. valida*.

Leptaena valentia Cocks, 1968
Pl. 12:1–4

Synonymy. – □1916 *Leptaena rhomboidalis* Wilck – Holtedahl (*pars*), pp. 69–72, Pl. 12:3, 6 *non* Pl. 12:1–2, 4–5, 7–12. □1968 *Leptaena valentia* sp. nov. – Cocks, p. 307, Pl. 8:1–8. □1968 *Leptaena valentia mullochensis* – Cocks, p. 309, Pl. 8:9–15. □1970 *Leptaena valentia* Cocks – Temple, p. 44, Pl. 12:1–10. □1978 *Leptaena valentia* Cocks, 1968 – Cocks, p. 117. □1987 *Leptaena haverfordensis* Bancroft, 1949 – Temple (*pars*), pp. 75–79, Pl. 8:2, 7. □1989 *Leptaena* cf. *valentia* Cocks, 1968 – Kul'kov & Severgina, pp. 136–137, Pl. 27:11.

Type specimen. – Holotype of *L. valentia*, by original designation; B 73340; conjoined valves from the Woodland Formation (Rhuddanian), Woodland Point, Girvan, Strathclyde. Grid ref. NX 168 952.

Material. – PMO 128.248, 128.378–128.381: internal moulds of 3 pedicle and 1 brachial valve and 1 external mould of a pedicle valve from the top of the Padda Member, Solvik Formation, at the west coast of Malmøya.

Remarks. – The present material is scarce and fragmentary. Measurements could be obtained from only two ventral valves (Table 38). A full description is therefore not given. The paucity of material is, however, due to insufficient collecting at Malmøya in the Oslo District. *Leptaena* is quite common at this locality, and since all material I collected there belong to this species, there is reason to believe that *L. valentia* is the dominant and possibly the only species present.

Temple (1987) noted that *L. valentia* was conspecific with *L. haverfordensis*. He based this assumption mainly on the configuration of the pedicle muscle scars, which he found to vary with the sagittal length of the disc. He also claimed that the general characters of the cardinalia in the two species were similar. There is certainly variation in the outline of the pedicle muscle scar. However, I find that there is a clear relationship between the configuration of the pedicle muscle scars and the different types of cardinalia. The evenly rounded pedicle muscle scars of *L. haverfordensis* occur with cardinalia having very elongate and widely spaced cardinal-process lobes, as seen in Pl. 11:3, 6 and 17 herein and Cocks (1968), Pl. 5:15, and Temple (1987), Pl. 8:7 and 10. The more parallel pedicle scars of *L. valentia* are paired with stout and

Table 38. Leptaena valentia (Cocks, 1968), measurements for ventral valves.

	s.l.	m.w.	h.w.
PMO 128.248 internal mould	17.00	43.00	43.00
PMO 128.379 external mould	–	22.00	22.00

closely spaced cardinal-process lobes, with a chilidium drawn over the lobes in a hood-like fashion and with a depression in the 'hood' between the lobes, as seen in Pl. 12:3 herein and Cocks (1968), Pl. 8:7, and Temple (1970), Pl. 12:3, 8, 9. I therefore regard *L. valentia* as a valid species.

Leptaena purpurea Cocks, 1968

Pl. 12:15

Type material. – By original designation; BB 31465; internal mould of pedicle valve, from Purple Shales (Telychian), Domas, Salop, Grid ref. SJ 594 006.

Synonymy. – □1968 *Leptaena purpurea* sp. nov. – Cocks, p. 313, Pl. 12:1–6. □1978 *Leptaena purpurea* Cocks – p. 116. □1982 *Leptaena purpurea* Cocks, 1968 – Cocks & Baarli, Pl. 1:10–15.

Remarks. – Only one pedicle valve (PMO 135.949) was collected from the Vik Formation at Sandvika. This species is common in the Vik Formation in other districts of the Oslo Region, and the pedicle valve seems to fall within the range of variation for this species.

Leptaena sp.

Pl. 12:10, 13

Remark. – This single brachial valve seems to be different from other *Leptaena* in this study. The rugae are very faint; it has an initially convex valve with a very pronounced 'shelf' before it geniculates sharply. The posterior muscle scars are fairly strongly impressed. This valve may belong to *L. haverfordensis*, since the cardinal-process lobes are similar; however, in addition to the above mentioned traits, the impressions of the muscle scars differs. The valve was found at Bærum Sykehjem near the top of the Leangen Member, Solvik Formation.

Crassitestella gen. nov.

Type species. – Here selected, the holotype of *Crassitestella reedi* (Cocks, 1968). Conjoined valves, B73341, from the Woodland Formation, Woodland Point, Girvan, Strathclyde (Reed 1917, Pl. 13:7; Cocks 1968, Pl. 10:1).

Derivation of name. – Latin *crassa*, thick, *testella*, diminutive of shell.

Diagnosis. – Thick-shelled, leptaenid genus with convex to rounded geniculate pedicle valve and sharply geniculate brachial valve. The ornament is like that of *Leptaena*, with ribs and continuous rugae, and the interior of the brachial valve possesses well-developed, outwardly concave transmuscle septa.

Remarks. – This, the only species known of the new genus, was originally described as *Leptaena* by Cocks (1968), but he noted that 'no other species of *Leptaena* resembles it' (p. 311). Both Cocks (1968) and Temple (1970) remarked on the resemblance of the internal brachial muscle field to *Cyphomena* Cooper 1956. Temple (1987) as a consequence moved the species to *Laevicyphomena* (Cocks, 1968), previously a subgenus of *Cyphomena*. *Laevicyphomena* is indeed very close, both in profiles and interiors, to the new genus; however, *Laevicyphomena* is defined as having a smooth valve. Temple (1987) chose to disregard the ornament. Ribbing in the Leptaeninae is very often used as the main criterion for definition of genera. I therefore erect the genus *Crassitestella* to embrace all leptaenid species with a convex to rounded geniculate pedicle valve and sharply geniculate brachial valve, a brachial valve with laterally concave transmuscle septa, and an ornament like that of *Leptaena*, e.g., strong ribs and well-developed continuous rugae.

It is possible that *Crassitestella*, *Laevicyphomena*, and *Cyphomenoidea* ought to remain subgenera of *Cyphomena*, as Cocks (1968) considered for the latter two taxa. They are all very close in internal characters. Temple (1987) pointed out, however, the closer relationship of *Laevicyphomena* (including *Crassitestella reedi*) to *Dactylogonia*, since that genus has laterally concave, paired ridges, in contrast to *Cyphomena* and *Cyphomenoidea* with laterally convex ridges. *Dactylogonia* may be distinguished from all three by its fine, parvicostellate ornament. All of them seem to belong to the same plexus as *Leptaena* and are here regarded as similar genera distinguished mainly by differences in ornament.

The new genus is closest to *Laevicyphomena* but can be distinguished by its rugose ornament, in contrast to the lack of ornament in the latter. The only species known, *Laevicyphomena feliciter* Cocks 1968, also seems to have peripheral flanges in the pedicle valve, as seen in some *Dactylogonia*, and a well-developed median ridge in the pedicle muscle field, both of which are absent in the present species.

Crassitestella reedi (Cocks, 1968)

Pl. 12:18, Pl. 13:1–11

Type material. – Same as for genus.

Synonymy. – □1917 *Leptaena rhomboidalis* (Wilckens) var. – Reed, p. 872, Pl. 13:7. □1968 *Leptaena reedi* sp. nov. – Cocks, p. 310, Pl. 10:1–14. □1970 '*Leptaena*' *reedi* Cocks, 1968 – Temple, p. 45, Pl. 12:11–18. □1978 *Leptaena? reedi* Cocks, 1968 – Cocks, p. 116. □1982 '*Leptaena*' *reedi* Cocks 1968 – Thomsen & Baarli, Pl. 2:8. □1986 '*Leptaena*' *reedi* Cocks – Baarli & Harper, Pl. 3f. □1987 *Laevicyphomena reedi* (Cocks, 1968) – Temple, pp. 80–81, Pl. 8:11–13.

Material. – PMO 105.203 (2 specimens), 128.233 (2 specimens), 128.236, 128.237 (2 specimens), 128.259, 128.260, 130.919–130.922: internal moulds of 3 pedicle and 5 brachial valves and 5 external moulds from the Myren Member, Solvik Formation, at Nesøya, Ostøya, Brønnøya and Konglungen, Asker.

Description. – The shell is thick with a subrectangular to trapezoidal outline. The maximum width occurs at the straight hinge line, which is produced laterally as ears. The valves are resupinate with a subrectangular to subcircular disc and length $\frac{1}{2}$–$\frac{3}{5}$ of maximum width. The pedicle valve is gently convex posteriorly, becoming flat before it geniculates in a rounded fashion at 70–110°. The brachial valve disc is flatter, with a sharper geniculation. The trail is long, up to equal the length of the disc, and there is often a slight indentation or fold on the trail. The ventral umbo is inconspicuous, with a low beak.

The pedicle foramen is round and open. A gently concave and apsacline ventral interarea has a wide delthyrium. The dorsal interarea is anacline, short and striated, with a notothyrium possessing small, convex chilidium plates extending laterally across the hinge line as raised plates with flanking grooves.

The disc has 6–8 regular rugae parallel to the disc periphery. The two outer rugae are most prominent, and the rugae become weaker towards the umbo. No rugae are developed on the trail. The radial ornament is multicostellate, with rounded, relatively high costae equally spaced at 4 per millimeter at 5 mm distance from umbo.

Interior of pedicle valve: The delthyrial cavity is broad and shallow. The teeth are strong, blunt and broad, situated in the plane of the hinge line. The dental plates are very short and stout, continuing as strong, slightly curved muscle-bounding ridges that never meet anteriorly. The muscle field is moderately to well impressed, occupying $\frac{1}{4}$–$\frac{1}{2}$ of the maximum length. The muscle scar shows a broad, deep, and raised indentation axially. The adductor scars are small, lanceolate to diamond-shaped on each side of a weak median septum. The two trunks of the vascula media diverge from the end of the median septum and delimit the diductor scars anteromedially. From the anterior end of the muscle scars, the trunks run parallel out to the periphery of the disc. The cordate diductor scars are broad and enclose the adductor scars. Sometimes the muscle field is drawn out to a delicate point anteriorly.

Interior of brachial valve: The strong cardinalia have a drop-shaped cardinal process in cross-section. The cardinal-process lobes are separated by a thin groove. The sockets are deep, subparallel with the hinge line, and may be crenulated with anteriorly splaying walls. The notothyrial platform is high and anchor-shaped, with strong socket ridges and a short median ridge developed in most specimens. The median ridge may be grooved and thus forms a double ridge. The muscle field is variable impressed, bearing paired, outwardly concave trans-muscle septa that may be very deep at mid-length of the valve and reach $\frac{1}{2}$–$\frac{3}{5}$ of the disc length. Another set of septa or muscle-bounding ridges may be developed posterolaterally to the trans-

Table 39. Crassitestella reedi (Cocks, 1968), measurements and statistics for the valves.

Ventral valves	s.l.	m.w.	h.w.
PMO 105.203 internal mould	6.00	11.00	11.00
PMO 128.237 internal mould	5.00	9.00	9.00

Dorsal valves	s.l.	m.w.	h.w.
Mean	6.20	11.20	11.20
Std. dev.	0.45	1.64	1.64
Count	5	5	5
Minimum	6.00	10.00	10.00
Maximum	7.00	14.00	14.00
Variance– covariance matrix	0.20	0.70 2.70	0.70 2.70 2.70

muscle septa. They extend from the lateral point of the socket ridges to the midpoint of the muscle field. The muscle scars are quadripartite with short, broad posterior scars and elongate, narrower anterior scars. A faint, low and broad axial ridge may reach as far anterior as the trans-muscle septa. There is no trace of rugae on the interior. The vascular mantle system is saccate.

Remarks. – Table 39 gives the measurement and statistics for *C. reedi*. This is a small species. There is no evidence of change in shape through growth, although the material is far too scarce to draw a definite conclusion.

This material is close to the type Llandovery material described by Cocks (1968) and Temple (1970, 1987). The Norwegian material, however, reveals some more morphological details, such as those of the mantle system.

Genus *Cyphomenoidea* Cocks, 1968

Type species. – By original designation *Leptaena wisgoriensis* Lamont & Gilbert, 1945, BU 382; a pedicle valve, the original of Lamont & Gilbert, 1945, Pl. 3:11, 12, from Wych Beds (Telychian), Coneygore Coppice, Alfrick, Worcestershire. Grid ref. SO 747 511.

Cyphomenoidea wisgoriensis (Lamont & Gilbert, 1945)

Pl. 13:12–23

Type material. – Same as type species.

Synonymy. – □1945 *Leptaena wisgoriensis* n.sp. – Lamont & Gilbert, p. 660, Pl. 3:10–14. □1968 *Cyphomena* (*Cyphomenoidea*) *wisgoriensis* (Lamont & Gilbert, 1945) – Cocks, p. 316, Pl. 12:9–12. □1978 *Cyphomena* (*Cyphomenoidea*) *wisgoriensis* (Lamont & Gilbert, 1945) – Cocks, p. 119. □1982 *Cyphomenoidea wisgoriensis* (Lamont & Gilbert, 1945) – Thomsen & Baarli, Pl. 3:9, 11, 12.

Material. – PMO 108.262, 108.272, 111.667 (2 specimens), 111.700, 117.414 (2 specimens), 130.924, 130.925 (2 specimens), 130.927–130.929, 130.931–130.934, 130.938–130.940, 130.941 (three specimens), 130.942, 130.943, 135.908, 135.921, 135.933, 135.947, 135.959, 135.966, 135.992: internal moulds of 9 pedicle and 18 brachial valves and 5 external moulds from the upper parts of the Rytteråker Formation at Kampebråten, Sandvika and near the base of the Vik Formation at Christian Skredsviks vei, Bærum.

Description. – Exterior: The shell is small and semi-rectangular, with the brachial valve concave and sharply geniculate, whereas the pedicle valve is evenly convex to a variable degree, without geniculation. The maximum width is at the straight hinge line that has small alae. The brachial valve geniculates at an 80–110° angle to produce a short trail that occupies ¼–⅓ of the total valve length. A few specimens have a raised edge around the periphery of the disc. The ventral beak bears a small but distinct foramen. The umbones are low. The ventral interarea is thin, straight, and apsacline. The delthyrium is wide, with angles of 90–110°. The dorsal interarea is also thin, flat and anacline. No chilidium is observed. The notothyrial walls have angles of 70–90°.

The ribbing on the disc is parvicostellate with 8–12 parvicostellae that continue on the trail. The costellae are very fine and regular, about 10 per millimeter at 5 mm from the umbo, and there are about 8 between each costa. Six to eight strong rugae are present over the entire disc. The rugae are offset by the parvicostellae, sometimes creating a distinct zig-zag pattern (Pl. 13:22). There are no rugae on the trail.

Interior of pedicle valve: The delthyrial cavity is small and relatively shallow, with a pedicle callist. The teeth are knob-like, bluntly rounded to subtriangular, and relatively large. The dental plates are strong but short, continuing in variably developed, strong ridges that are evenly rounded towards the median ridge. The median ridge is broad, sometimes high, and widening anteriorly.

The muscle field is deeply impressed, occupying ⅕–⅓ of the maximum width. It consists of a pair of broadly triangular diductor scars and long, lanceolate adductor scars that are situated on each side of the median ridge and surpassing the diductor scars anteriorly. The extra-muscular areas are impressed by rugae.

Interior of brachial valve: The cardinalia are strong and situated on a high notothyrial platform. The platform extends anteriorly as a high, relatively short, and broad median ridge, ⅓–⅖ of the total disc length. The ridge bifurcates at a slight angle anteriorly. The cardinal-process lobes are small, facing ventrally to posteroventrally. They are parallel to each other and have a deep, narrow groove between them. The socket ridges are thick, high and relatively long with an angle of about 20° to the hinge line. The sockets are shallow and elongate but bordered by the high socket ridges on the anterior side. The muscle fields are deeply impressed, up to ⅖ of the total disc width, and ⅘ of the total disc length. The posterior pair is partly lateral to the anterior pair and is rounded flabellate in shape. The posterior muscle scars are bounded posteriorly by the socket ridges and divided by the median ridge. A pair of posterolaterally extending trans-muscle septa may divide the posterior muscle scar in two, with the smaller parts posterolaterally. The anterior muscle scars are elongate and bordered by a pair of long, deeply impressed ridges that may curve anteromedially. A thin, strong median ridge occurs towards the anterior end of the anterior muscle-bounding ridges. This ridge often bifurcates and reaches a position adjacent to or a little farther anteriorly than the bounding ridges.

Remarks. – This is a small species (Tables 40, 41). The covariance between sagittal length and maximum width is extremely low, especially for the ventral valve. Inspection of the valves and the data (Fig. 11) shows a great plasticity in shape. This also includes the depth of the valves. The regression lines for dorsal and ventral valves vary widely, which may be explained by the predominance of plots in a narrow length interval and by variation in shape for the few deviating length plots for dorsal and ventral valves. Potential allogenic growth may be masked by the variation within each growth stage.

The Norwegian material is close to the British type material of this species but may differ in having thicker socket ridges. The Norwegian material is from a slightly older horizon than the British material, but they are both Telychian in age and certainly close enough morphologically to belong to the same species.

Table 40. Cyphomenoidea wisgoriensis (Cocks, 1968) measurements and statistics for the ventral valves.

	s.l.	m.w.	h.w.
Mean	5.88	8.25	8.17
Std. dev.	0.79	1.23	1.46
Count	8	10	9
Minimum	5.00	7.00	6.00
Maximum	7.00	10.00	10.00
Variance–	0.62	0.26	0.37
covariance		1.51	1.85
matrix			2.13

Table 41. Cyphomenoidea wisgoriensis (Lamont & Gilbert, 1945), measurements and statistics for the dorsal valves.

	s.l.	m.w.	h.w.
Mean	4.89	7.74	7.74
Std. dev.	0.84	1.45	1.45
Count	19	19	19
Minimum	3.00	5.00	5.00
Maximum	6.50	10.00	10.00
Variance–	0.71	0.75	0.75
covariance		2.09	2.09
matrix			2.09

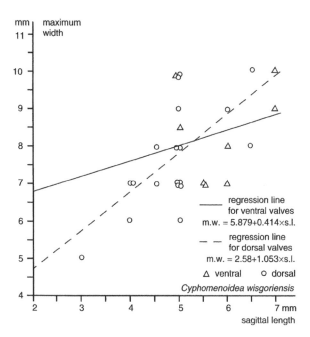

Fig. 11. Scattergram with regression lines for ventral and dorsal valves of *Cyphomenoidea wisgoriensis* (Lamont & Gilbert, 1945), plotting sagittal length against maximum width.

Family Leptostrophiidae Caster, 1939

Genus *Eostropheodonta* Bancroft, 1949

Type species. – Original designation, *Orthis hirnantensis* Mc-Coy, 1851, p. 395, the Upper Ordovician (Hirnantian) of Bala, Wales.

Eostropheodonta multiradiata? Bancroft, 1949

Pls. 13:24–26; 14:1–8

Type material. – Designated by Cocks (1978, p. 126). Internal and external moulds of ventral valve, A30113a, b, from the Millin Mudstone Formation (Aeronian), Fortune's Frolic, Haverfordwest, Dyfed. Grid ref. SM 967 143. Figured by Bancroft (1949, Pls. 2:8; 9:1).

Synonymy. – □1916 *Rafinesquina expansa* – Holtedahl, p. 26, Pl. 2:6, 7 *non* Pl. 2:8–13. □1949 *Eostropheodonta multiradiata* sp. nov. – Bancroft, p. 10, Pl. 2:8. □1967 *Eostropheodonta multiradiata* – Cocks, p. 255. □1978 *Eostropheodonta multiradiata* – Cocks, p. 126. □1982 *Eostropheodonta* sp. – Thomsen & Baarli, Pl. 2:4, 6. □1987 *Eostropheodonta* sp. – Baarli, Fig. 6b. □1987 *Eostropheodonta multiradiata* – Temple, pp. 81–83, Pl. 9:1–9.

Material. – PMO 103.466, 103.467 (4 specimens), 108.274, 128.392, 128.397, 128.401, 130.328, 130.329, 130.330 (3

specimens), 130.331, 130.332 (2 specimens), 130.334, 130.891: internal moulds of 9 pedicle and 3 brachial valves and 7 valve exteriors from the Leangen Member, Solvik Formation, at Skytterveien, Asker. This species is also found in the Leangen Member at Bleikerveien, Asker, and at Jongsåsveien, Sandvika.

Description. – Exterior: Planoconvex to slightly biconvex valves with low ventral curvature and semicircular outline, $\frac{3}{4}$–$\frac{4}{5}$ as long as wide. The maximum width is at the hinge line, which is mostly alate and may be sharply mucronate. The lateral and ventral margins are evenly rounded, with a rectimarginate commissure. The umbo is low and the beak erect, barely protruding beyond the hinge line. The ventral interarea is low, straight and apsacline, with a wide, open delthyrium. The dorsal interarea is anacline, and the chilidium seems to be lacking, but that could be due to preservation.

The ornament is equally parvicostellate to nearly multi-costellate. The costae are rounded and straight-sided, with 2–4 lower costellae between the primary costae. There is about 1 costa per millimeter in specimens that are clearly parvicostellate, and about 4 costellae per millimeter in the more multicostellate specimens. The ribs are straight centrally and curve gently laterally. The costellae originate by branching, and faint rugae may be present near the posterior margin.

Interior of pedicle valve: The delthyrial cavity is shallow and occupied posteromedially by a small, rounded ventral process. There is no myophragm. The dental plates are well developed, relatively short, concave outward and diverging at 60–80°. A pair of low, short denticular plates is fused with the dental plates at the hinge line. The dental plates may continue as short to vestigial muscle-bounding ridges. The muscle field is indistinct. One specimen (Pl. 14:2) shows thin and long, lanceolate adductor muscle scars on each side of a thin median ridge and two radial ridges on each side across the diductor scars. The extra-muscular area is coarsely tuber-culous and often impressed by external ornament.

Interior of brachial valve: The notothyrial platform is low, but distinct, and continuing in a broad median ridge reaching $\frac{1}{4}$–$\frac{1}{3}$ of the total valve length before it disappears. The cardinal-process lobes are oblong, tapering posteriorly and facing ventroposteriorly. They diverge at about 30–40°. There is a thin ridge between the lobes near the hinge line. The thick, high socket plates continue in front of the cardi-nal-process lobes. They have strong denticles on the poste-rior edges, which are undercut. The socket plates diverge by about 130°. There are slight thickenings of the hinge line on each side of the cardinal-process lobes. The extra-muscular area is coarsely tuberculous. The muscle fields are not im-pressed, but the non-tuberculous areas indicate oval, rela-tively small fields.

Remarks – This is a medium sized species (Table 42). The covariance between sagittal length and maximum width is high, although the material is very limited.

Table 42. Eostropheodonta multiradiata? (Bancroft, 1949), measurements and statistics of valves.

Ventral valves	s.l.	m.w.	h.w.
Mean	15.78	21.67	21.33
Std. dev.	5.29	6.06	5.75
Count	9	6	6
Minimum	6.00	10.00	10.66
Maximum	24.00	27.00	25.00
Variance–	27.94	32.00	30.20
covariance		36.67	34.53
matrix			33.07

Dorsal valves	s.l.	m.w.	h.w.
PMO 103.466 internal mould	9.00	–	–
PMO 130.330 internal mould	5.00	8.00	8.00

This material is similar to several species defined from Great Britain. It is probably closest to *E. multiradiata* Bancroft, 1949, but differs in its more multicostellate ribbing, smaller size, less transverse outline and more divergent cardinal-process lobes. Temple (1987), however, claimed that smaller forms had a less transverse outline, while Rong & Cocks (1994) observed that ribbing is very variable in this genus and of limited taxonomic use. It differs from *E. vogariensis* Cocks, 1967, in less divergent dental plates and less divergent sockets and in having a median ridge in the dorsal valve. It differs from *E. mullochiensis* (Reed, 1917) in having less pronounced parvicostellate and more regular ornament and a relatively larger cardinalia. All these species are very close, and differences are small. However, these species are not very plastic in morphology. I therefore only tentatively place the Norwegian form with *E. multiradiata* and feel that later it might be justified to erect a new species, given closer investigation and larger material of the two populations.

Eostropheodonta delicata sp. nov.
Pl. 14:9–15

Holotype. – A brachial valve, PMO 128.376, Pl. 14:10, 14, from 32 m above the base of the Leangen Member, Solvik Formation, Leangbukta, Asker.

Derivation of name. – Latin *delicatus*, dainty, with regards to the cardinalia.

Material. – PMO 103.470, 103.471, 103.474, 103.510, 103.572, 128.376, 128.377, 128.385 (two specimens), 130.342, 130.364: internal moulds of 4 pedicle and 2 brachial valves and 5 external moulds from the Leangen Member, Solvik Formation, at Leangbukta and Skytterveien, Asker.

Diagnosis. – This is a small, consistently multicostellate *Eostropheodonta* species with thin shells and very short, widely diverging dental plates. The cardinalia are very small and delicate compared to valve size, and the notothyrial platform is indistinct.

Description. – Exterior: Planoconvex to concavo-convex, very thin and small shells that are transversely subcircular and $\frac{2}{3}$ as long as wide. The maximum width is at the hinge line, which is extended as sharply pointed alae. The lateral margins are gently rounded, while the anterior margin is evenly rounded and rectimarginate. The umbo is weak and does not protrude over the hinge line. The dorsal interarea is very short and straight, while the ventral interarea is larger and apsacline.

The ornament is finely and equally costellate with about 7 low and thin costae per millimeter. The costae are somewhat irregular in width, and new costae arise by branching. Rugae are common near the hinge line.

Interior of pedicle valve: The delthyrial chamber is shallow, with a very small ventral process. The dental plates are very short, low, and diverge at 110–120°. There is generally no myophragm and no muscle-bounding ridges. The muscle scars are often overprinted by ribbing. One specimen, PMO 103.510, shows minute adductor scars enclosed by small oblong diductor scars that are divided by a very weak ridge and bisected by faint radial ridges.

Interior of brachial valve: The cardinalia are very delicate and minute. There is no clearly developed notothyrial platform or median ridge. The cardinal-process lobes are thin and face ventrally to posteroventrally with bases diverging anterolaterally. The angle of divergence is about 50°. There may be a very thin ridge medially between the cardinal-process lobes. The socket plates are very low, straight, and somewhat longer than the cardinal-process lobes. They diverge about 30° to the hinge line and have 7–8 denticles on their posterior faces. The sockets are low. There is no trace of the muscle scars and the ribbing is overprinted over the entire interior of the valve.

Remarks. – The material is scarce, but the available measurements are presented in Table 43. The ribbing is close to that of *E. vogariensis* Cocks, 1967, but the new species differs in its generally smaller size, minute cardinalia, and much shorter dental plates with a wider angle. The same characteristics distinguish it from the Norwegian populations of *E. multiradiata* together with the finer and consistently multicostel-

Table 43. Eostropheodonta delicata sp. nov., measurements and statistics for the valves.

Ventral valves	s.l.	m.w.	h.w.
PMO 103.471 internal mould	–	11.00	11.00
PMO 103.404 external mould	–	13.00	13.00
PMO 103.510 internal mould	15.00	15.00	15.00
PMO 130.342 internal mould	–	21.00	21.00
PMO 130.364 (parat.) int. mould	14.00	18.00	18.00

Dorsal valves	s.l.	m.w.	h.w.
PMO 103.470 internal mould	–	12.00	12.00
PMO 128.376 (holot.) int. mould	–	18.00	18.00

late ribbing. There is, however, a considerable overlap in size range between the two species, where the difference in the size of cardinalia is striking. Evenly costate ribbing in *Eostropheodonta* is rare and no other species seems to be close.

Genus *Palaeoleptostrophia* Rong & Cocks, 1994

Type species. – *Stropheodonta jamesoni* Reed, 1917, selected by Cocks (1978, p. 127), B73036, the original of Reed 1917, Pl. 16:29, from the Woodland Formation (Rhuddanian), Woodland Point, near Girvan, Scotland.

Palaeoleptostrophia ostrina? (Cocks, 1967)
Pl. 15:1–6, 8, 11, 14

Type species. – By original designation. A brachial valve, OUM C14014, from the Purple Shales of Hughley Brook (Grid ref. SO 5605 9747).

Synonymy. – □1967 *Leptostrophia* (*Leptostrophia*) *ostrina* Cocks, 1967, p. 252, Pls. 37:12–14; 38:1. □1994 *Palaeoleptostrophia ostrina* (Cocks, 1967) – Rong & Cocks, p. 684.

Material. – PMO 103.477, 128.321, 128.322, 130.324, 130.326, 130.327, 130.340, 130.344: internal moulds of 3 pedicle and 3 brachial valves and 1 valve exterior from the Leangen Member, Solvik Formation, at Leangbukta and Skytterveien, Asker.

Description. – Exterior: Large, gently concavo-convex valves that have a broadly subcircular outline about $3/5$ as long as wide. The maximum width is at the straight hinge line. The hinge line may be extended as slight alae or be broadly rounded. The lateral margins are evenly rounded, and the anterior margins slightly rounded to nearly straight. The commissure is rectimarginate. The umbones are very low. The ventral interarea is apsacline, low and plane. The delthyrium is open or may possibly have a small pseudodeltidium apically. The delthyrial angles are very wide, diverging at about 120°. The dorsal interarea is straight and anacline, and there are no details on the notothyrium.

Both the external valve and the internal moulds have about 15–17 costae per 10 mm at 10 mm from the umbo. The secondary costellae are rounded and unequal in size, often with the highest and thickest costellae in the middle between the two primary costae. There are 7–9 costellae between each costa. Rugae are present posterolaterally.

Interior of pedicle valve: The delthyrial chamber is broad and shallow, with a weak ventral process that tapers into a low, thin myophragm, which continues for about $1/3$ of the total valve length. The dental plates are absent or rudimentary; instead there are high, short, thick and triangular dental lamellae with 5–6 denticles each. They occupy less than $1/4$ of the total valve length. The muscle field is broadly triangular and very weakly impressed. It may be bounded posterolaterally by weak, straight ridges. The angle of divergence is 90–100°. The adductor scars are relatively long and situated on each side of a myophragm. The diductor scars continue to the anterior end of the myophragm and are possibly flabellate. The extra scar areas are finely tuberculate in rows along costae, which often are lightly impressed.

Interior of brachial valve: The notothyrial platform is low to near absent. It may be extended anteriorly in a broad, long and tapering median ridge, occupying half of the total valve length. The cardinal-process lobes are oblong, relatively stout, and directed anterolaterally. They diverge at about 45°, while their bases are directed more anteriorly. The sockets are shallow. Thin socket ridges are about the same length as the processes, and they get thicker as they extend anterolaterally at an angle of 30–40° to the hinge line. About $1/3$ of the hinge line is denticulated with small denticles. The muscle field is feebly impressed but probably oblong and occupies $1/2$ of the valve length.

Remarks. – Table 44 shows that this is a large species. The covariance of the sagittal length and maximum width has a negative value, but this is based on only 4 valves. The species is tentatively assigned to *P. ostrina*. Although inadequately described originally, the transverse shape, the weakly impressed, broad muscle scar, and the ornament are similar in the two samples. The type material of *P. ostrina* differs, however, in its smaller size and allometric growth pattern. The allometric growth described by Cocks (1967) lead to near equal length and width in large valves in the type material, but all of the present material is considerably wider than long. *P. ostrina* occurs in the Purple Shales (Telychian) of Shropshire, England and is thus younger than the Norwegian material from lower Aeronian strata.

Moreover, most of the brachial valves included in *P. ostrina?* are very small compared with the pedicle valves, and these small valves may belong to another taxon, as further explained below under *?Eomegastrophia* spp?.

Table 44. Palaeoleptostrophia ostrina? (Cocks, 1967), measurements and stratistics for the ventral valves. No measurements were obtained for dorsal valves.

	s.l.	m.w.	h.w.
Mean	26.50	43.25	43.25
Std. dev.	2.52	2.75	2.75
Count	4	4	4
Minimum	24.00	40.00	40.00
Maximum	30.00	46.00	46.00
Variance–	6.33	3.50	3.50
covariance		7.58	7.58
matrix			7.58

Genus *Mesoleptostrophia* Harper & Boucot, 1978

Type species. – *Mesoleptostrophia* (*Mesoleptostrophia*) *karta-lensis* Harper & Boucot, 1978a, Pl. 2:1–5, by original designation, from the Lower Devonian of Turkey.

?*Mesoleptostrophia* sp.
Pl. 15:9, 12, 15

Material. – PMO 135.933, 135.938, 135.943: One pedicle valve and 2 brachial valves, from the uppermost parts of the Rytteråker Formation.

Remarks. – The rare and fragmentary nature of the material prevents a firm assignment to a taxon. If the dorsal and ventral valves belong to the same species, *Mesoleptostrophia* is a likely genus.

Genus *Eomegastrophia* Cocks, 1967

Type species. – By original designation; GMS 10215; a pedicle valve, from Pentamerus Beds (Aeronian), Morellswood, Salop. Grid. Ref. SJ 628 064.

Eomegastrophia spp.?
Pl. 16:7–12

Synonymy. – □1916 *Brachyprion* sp. or *Rafinesquina* sp. – Holtedahl, Pl. 5:5.

Material. – PMO 103.546, 103.638, 128.391, 128.393, 128.398, 128.400, 130.325, 130.337: internal moulds of 6 pedicle and 2 brachial valves from the Leangen Member, Solvik Formation, at Skytterveien and Leangbukta, Asker.

Description. – Exterior: Medium-sized to large valves, semicircular and about equally as long as wide. The shells are very thin, with pedicle valves of varying convexity, from gently convex in the smaller valves to strongly convex in the larger. The brachial valves are gently concave. The hinge line is straight and coincides with maximum width. It may be extended as broadly rounded alae. The cardinal angles are rounded, with lateral and anterior margins evenly curved. The commissure is smooth, and the anterior commissure is rectimarginate.

The beak is relatively small and erect, extending slightly posterior of the hinge line. The ventral interarea is flat, very low, and apsacline to orthocline. The delthyrium is sealed by a concave pseudodeltidium. The dorsal interarea is very low, and there is no information on the notothyrium.

The ornament is very fine and often not impressed internally. The costae are irregular, with a tendency towards parvicostellate ribbing. There are 6 costae per millimeter at 10 mm from the umbo. Some specimens have greater differentiation towards parvicostellation than others. The costae are low and rounded.

Interior of pedicle valve: The dental plates are either very short or absent altogether. When present, they diverge at a highly variable angle from 90° to 145°. Short denticular plates possess 6–8 small denticles. The muscle field is large, $^2/_5$ of the total length and $^1/_2$ of the total width, and oval to subtriangular. Low muscle-bounding ridges may be developed posteriorly. The muscle field is faintly impressed and open anteriorly. The diductor scars are longitudinal, oval, and enclose the adductor scars laterally. The adductor scars are situated on each side of a thin to negligible median ridge. They are relatively small and lanceolate. The median ridge may extend to the anterior part of the diductor scars. The extra muscular areas of the valves are coarsely tuberculate.

Interior of brachial valve: The notothyrial platform is weakly developed. Straight, discrete brachiophores diverge at 60–70°. Their faces are relatively thin and directed ventrally, with a small, thin ridge posteriorly between them. The socket ridges are very low and curved, ending up being subparallel to the hinge line. There are a few denticles close to the umbo. The notothyrial platform continues anteriorly in a weak median ridge, which may reach half-way to the anterior margin. The muscle fields are obscure but possibly semicircular. There seem to be two very faint ridges bisecting the muscle scars. The ridges arise just anterior of the notothyrial platform.

Remarks. – Table 45 shows a large variation it size, in spite of very limited material. *Eomegastrophia* is characterized by the presence of dental plates, while *E.* spp.? shows specimens both with and without this trait. The material may thus belong to several taxa. The closeness in convexity, denticular plates and shape of muscle scars, however, suggest they all belong to the same species, at least the larger specimens. The British *Eomegastrophia ethica* Cocks, 1967, is very similar to the specimens with dental plates. The only other difference

Table 45. Eomegastrophia spp.?, measurements and statistics for valves.

Ventral valves	s.l.	m.w.	h.w.
Mean	23.17	22.50	22.50
Std. dev.	6.31	9.26	9.26
Count	6	4	4
Minimum	13.00	9.00	9.00
Maximum	29.00	30.00	30.00
Variance–	39.77	67.50	67.50
covariance		85.67	85.67
matrix			85.67

Dorsal valves	s.l.	m.w.	h.w.
PMO 128.400 internal mould	–	28.00	28.00
PMO 130.325 internal mould	15.00	20.00	20.00

observed here is the lack of an alveolus in front of the cardinal process lobes. The small pedicle valves are difficult to distinguish from *Eopholidostrophia*, but the latter is more transverse and has a more triangular muscle field. Some of the brachial valves assigned to *Palaeoleptostrophia ostrina*? may belong to *Eomegastrophia* spp.?, but placement is on the basis of denticulation. Those specimens with the greater part of the hinge line denticulated, stronger socket ridges and less divergent cardinal-process lobes, were placed with *P. ostrina*?

Family Eopholidostrophiidae Rong & Cocks, 1994

Genus *Mesopholidostrophia* Williams, 1950

Type species. – By original designation, *Pholidostrophia (Mesopholidostrophia) nitens* Williams, 1950; from the Mulde Beds (Wenlock) of Gotland, Sweden, a junior subjective synonym of *Leptaena laevigata* J. de C. Sowerby, 1839.

Mesopholidostrophia sifae sp. nov.

Pl. 15:7, 10, 13,16. Pl. 16:1–6.

Synonymy. – □1982 *Brachyprion arenacea* (Davidson) – Cocks & Baarli, Pl. 3:10, 13, 14.

Holotype. – PMO 135.932; Internal mould of pedicle valve, found 80 m above the base of the Rytteråker Formation, Kampebråten, Sandvika.

Derivation of name. – From the Norse goddess Sif.

Material. – PMO 108.266, 108.268, 111.414, 111.696, 117.411 (2 specimens), 130.933, 130.938, 130.941, 135.906, 135.907, 135.913, 135.915 (2 specimens), 135.916, 135.917, 135.918 (2 specimens), 135.919, 135.921–135.927, 135.932, 135.936–135.938, 135.941, 135.987: internal moulds of 26 pedicle and 7 brachial valves from the uppermost parts of Rytteråker Formation at Kampebråten and around the base of the Vik Formation at Christian Skredsvik vei in Bærum.

Diagnosis. – Medium-sized *Mesopholidostrophia* with very variable convexity and weak parvicostellate ornament. The chilidium is small and the delthyrium is open. It has weakly impressed diductor scars and a small process with process pits continuing in a well-developed median ridge in the ventral valve. The oval, dorsal muscle scars are strongly impressed and the small socket ridges are straight and widely spread.

Description. – Exterior: There are all intergradations between gently and strongly convex pedicle valves. The brachial valves are gently concave to nearly flat. The convexity in the pedicle valve is greatest posteriorly, while it flattens anteriorly. The valves are semicircular, $\frac{3}{5}$ to equally as long as wide,

although with predominance of transverse valves. The hinge line is straight and often extended laterally into small alae. The maximum width is at the hinge line or, more rarely, in the posterior $\frac{1}{3}$ of the valve. The lateral margins are evenly rounded to subparallel, whereas the anterior margin is evenly rounded. The commissure is smooth and rectimarginate. The dorsal umbo and beak are weak, while the ventral umbo may be swollen and slightly overhangs the hinge line. A relatively long interarea is flat and orthocline to apsacline. An open delthyrium has delthyrial angles of 60–80°. A chilidium seems to be present in a straight and relatively long dorsal interarea.

The radial ornament is very weak and parvicostellate with about 1 thin and subangular costa per millimeter at the 10 mm growth stage from the umbo. The space between the costae may vary, as does the number of costellae that are often seen as weak threads. The number of costellae between each costa varies between 3 and 8, with an average of 5.4 on the two specimens where costellae were preserved.

Internal of pedicle valve: The ventral process is small and short, produced anteriorly as a variably developed myophragm. The myophragm is always present posteriorly, may continue for $\frac{2}{5}$ of the total length, and it is relatively high posteriorly. The process pits are short and shallow. Very short and protruding denticular plates may occupy the hinge line or may be variably fused to the hinge line. The denticulation occupies from $\frac{1}{2}$ to less than $\frac{1}{4}$ of the total width. The denticles are very small and rounded. Weak to absent muscle-bounding ridges diverge, where present, at 70–90°. The muscle scars are always open anteriorly. The indistinct adductor scars are lanceolate in shape on each side of the myophragm. The diductor scars, which are generally not impressed, seem to be broadly suboval, reaching less than $\frac{1}{2}$ the length of the valve.

Interior of brachial valve: The cardinalia are raised on a well-developed, anchor-shaped platform. The cardinal-process lobes are stout, elongate, and diverging anteriorly with faces directed ventrally. There may be a very fine ridge between the lobes near the hinge line. The sockets are shallow, with relatively thin, low, and short socket ridges, diverging from the hinge line at about 20–30°. The notothyrial platform has anterolaterally directed continuations or ridges that delimit the posterior parts of the well impressed adductor scars. Medially there is a thick, broad, median ridge that continues anteriorly as a breviseptum. There may be two very weak brachial ridges diverging on each side of the breviseptum and thus bisecting the muscle scars that become indistinct anteriorly. The entire valve with the exception of the notothyrial platform and ridges, is tuberculate.

Remarks. – Table 46 gives the measurements and statistics for *M. sifae*. Only the ventral valves gave adequate material for measurements. The covariance between the sagittal length and maximum width is slightly off. Fig. 12 shows that the variation within each growth stage is limited but there is a certain degree of allometric growth, the small valves being relatively wider than the larger ones.

Table 46. Mesopholidostrophia sifae sp. nov., measurements and statistics of valves.

Ventral valves	s.l.	m.w.	h.w.
Mean	14.53	18.39	18.06
Std. dev.	5.40	5.95	6.05
Count	17	18	17
Minimum	5.00	10.00	10.00
Maximum	27.00	34.00	34.00
Variance–	29.14	30.40	32.88
covariance		35.43	36.54
matrix			36.56

Dorsal valves	s.l.	m.w.	h.w.
PMO 111.414 internal mould	–	13.00	13.00
PMO 130.933 internal mould	–	16.00	16.00
PMO 135.906 internal mould	–	21.00	21.00
PMO 135.936 internal mould	–	14.00	14.00
PMO 135.938 internal mould	–	11.00	11.00

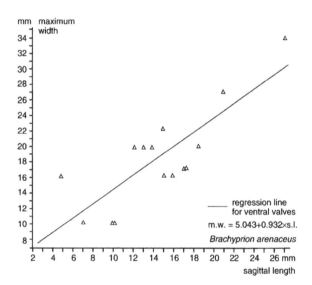

Fig. 12. Scattergram with regression line for ventral valve of *Mesopholidostrophia sifae* sp. nov., plotting sagittal length against maximum width.

Hurst (1974) reviewed the Eostrophiidae of Britain and found that generally there is a progressive increase in the size of the denticular plates and the impression of the ventral muscle field through time. *Mesopholidostrophia sifae* sp. nov. occurs in the latest Llandovery, contemporary to *M. salopiensis salopiensis* Cocks, 1967. The two agree well in the characters mentioned above, with the denticular plates occupying $^1/_4$–$^1/_2$ of the hinge line and with variably impressed adductors but obscure diductors. In size, shape and ornament, however, *M. salopiensis salopiensis* is closer to the Llandovery *Eopholidostrophia* stocks, while *M. sifae* is a more typical *Mesopholidostrophia* and close to the flatter and larger Wenlock and Ludlow species. In addition, it differs from *M.*

salopiensis salopiensis in having straight, not curving, socket plates and a smaller chilidium.

Cocks & Baarli (1982) formerly placed this material in synonymy with *Brachyprion arenaceus* (Davidson, 1871), and the two are much alike. Recently, Rong & Cocks (1994) revised *Brachyprion*. Clearly neither of the species above suits in the revised genus *Brachyprion*, because they both posess well-developed ventral processes, and especially *M. sifae* has a well-impressed dorsal muscle field. The latter is not seen in the type species *B. leda* Billings, 1860, and is uncommon in *Brachyprion*, according to Harper & Boucot (1978c). *M. sifae* also differs from *B. arenaceus* in possessing a better-developed notothyrial platform but weaker socket ridges and weaker ornament, which all are traits common to *Mesopholidostrophia*. *M. sifae* has more secondary costella between costae than *B. arenaceus*; 3–8 in the fomer compared to 2–4 in the latter.

Genus *Eopholidostrophia* Harper, Johnson & Boucot, 1967

Type species. – *Stropheodonta* (*Brachyprion*) *sefinensis* Williams, 1951; SM A30051; a pedicle valve from Aeronian beds on the bank of river Sefin, Llandovery, Dyfed. Grid ref. SN 742 282.

Eopholidostrophia spp.
Pl. 16:13–15

Material. – PMO 105.228, 105.229, 130.354: internal moulds of 3 pedicle valves from the Myren Member at Spirodden and the Leangen Member at Skytterveien, all the Solvik Formation, Asker.

Remarks. – This is a rare taxon, found both in strata of Rhuddanian and Aeronian age. The specimens from the uppermost parts of the Leangen Member are close to, and – judged from the parvicostellate ribbing and accentuated midrib – probably belong to, *E. sefinensis* Williams, 1951. As no brachial valves have been recovered, I defer from assigning them to a named species. The measurements for the available material are given in Table 47.

Table 47. Eopholidostrophia spp., measurements for ventral valves. No dorsal valves were recovered.

	s.l.	m.w.	h.w.
PMO 105.228 internal mould	7.00	12.00	12.00
PMO 105.229 internal mould	5.00	9.00	9.00
PMO 130.354 internal mould	11.00	13.00	13.00

Family Amphistrophiidae Harper, 1973

Subfamily Mesodouvillininae Harper & Boucot, 1978

Genus *Eocymostrophia* gen. nov.

Type species. – *Eocymostrophia balderi* from the base of the Vik Formation (Telychian), Kampebråten, Sandvika.

Derivation of name. – Greek *eos*, early. This is the earliest known species in the lineage leading to *Cymostrophia*.

Diagnosis. – The valves are gently concavo-convex, with a parvicostellate and rugose ornament where the primary costae interrupt the rugae. Denticles are set on short denticular plates. Well-developed trans-muscle ridges and breviseptum are present in the dorsal valve. The presence of denticular plates distinguishes it from *Mesodouvillina*.

Remark. – This genus is possibly the oldest representative of its subfamily, Mesodouvillininae, being of late Llandovery age. The species *Mesodouviella* sp. from the French River Formation of Nova Scotia, described by Harper (1973, pp. 41, 91–95), may be of the same age or slightly younger. The presence of denticular plates in *Eocymostrophia*, which is a primitive trait, indicates a position close to the origin of the Mesodouvillininae.

Mesodouviella Harper & Boucot, 1978, with its gentle concavo-convex outline, is thought by Harper & Boucot (1978b) to give origin to the more strongly concavo-convex genus *Mesodouvillina* Williams 1950. Later *Mesodouvillina* led to *Protocymostrophia* Harper & Boucot, 1978, through developement of parvicostellate ribbing broken by rugae. *Eocymostrophia* gen. nov., of the same age as or older than *Mesodouviella* shows, however, direct affinity with the *Protocymostrophia* lineage in having trans-muscle ridges that are variably but sometimes stronger developed than in other Mesodouvillininae and in having the characteristic ornament of rugae broken by ribs. The shape is more like that of *Mesodouviella*. Stronger curvature must be a trait acquired repeatedly in different lineages. In conclusion, *Mesodouviella* to *Mesodouvillina* and *Eocymostrophia* to *Protocymostrophia* must be parts of two closely related lineages that evolved in parallel.

Eocymostrophia balderi gen. and sp. nov.
Pl. 17:1–10, 13

Holotype. – PMO 135.968 (Pl. 17:7) a brachial valve with external counterpart PMO 135.967 (Pl. 17:13) from the base of the Vik Formation, Kampebråten, Sandvika.

Derivation of name. – Named after the norse God Balder.

Material. – PMO 135.935, 135.939, 135.942, 135.944, 135.945, 135.956, 135.967, 135.968, 135.970: internal moulds of 3 pedicle and 4 brachial valves and 2 external moulds of valves from the uppermost parts of the Rytteråker Formation at Kampebråten, Asker.

Diagnosis. – The valves are transverse, with variably developed muscle-bounding ridges that surround a cordate ventral muscle field posteroventrally. A ventral myophragm is present together with a well-developed ventral process with strong pits. The dorsal valve has strong cardinal-process lobes that generally converge anteriorly into a Y-like structure. The socket ridges are very low and widely divergent. A well-developed pair of dorsal trans-muscle ridges may be accompanied by a second faint and lateral pair.

Description. – Exterior: The large, weak to moderately concavo-convex shell has a semicircular outline, $\frac{3}{5}$ as long as wide. The hinge line is straight, and maximum width occurs between the valve midlength and the hinge line. The cardinal angles are broadly rounded into evenly rounded lateral margins, while the anterior margins seems to be slightly rounded. The umbones are low, with the ventral beak barely protruding beyond the hinge line. The ventral interarea is fairly high, straight, and apsacline. The delthyrium may have a tiny pseudodeltidium while a well-developed ventral process fills the apex. The delthyrial angles are very wide 130–150°. The dorsal interarea is flat and anacline. There is no information on the chilidium.

The ornament is regular and very finely parvicostellate with about 7 high, narrow costae per 10 mm at 10 mm from the umbo. There are 9–10 very low, rounded costellae between each costa. Fine rugae broken by primary costae are superimposed over the entire valve. The costae and costellae are straight medially and curve gently laterally.

Interior of pedicle valve: The delthyrial cavity is relatively shallow and occupied posteriorly by a broad, pronounced ventral process that, combined with strong ventral-process pits, give an arrow-like appearance. Posteriorly the process is divided by a very thin groove that continues into a thin myophragm. The myophragm runs to the anterior margin of the cordate muscle field. Dental plates are absent, but there are well developed denticular plates on the hinge line, occupying $\frac{1}{4}$ of the total length and displaying 7–8 denticles. The muscle field is broadly cordate and relatively well impressed. The muscle-bounding ridges vary from being short and straight, situated posterolaterally, to being longer and curving to encompass the thicker parts of the heart shape. They do not, apparently, delimit the muscle field anteriorly. The angle of the muscle-bounding ridges, posteriorly, is about 75°. Since there are no pedicle valves with the entire hinge line and margins preserved, dimensions are hard to estimate, but relative to the available brachial valves, the width of the muscle field would be $\frac{1}{3}$ or more of the total width and less than $\frac{1}{2}$ of the total length. The adductor scars are oblong and relatively small. They are divided by the myophragm in their anterior half and surrounded by large drop-shaped diductor scars. One of the specimens has two pairs of very weak radial

Table 48. Eocymostrophia balderi sp. nov., measurements of valves.

Dorsal valves	s.l.	m.w.	h.w.
PMO 135.935 external mould	22.00	40.00	32.00
PMO 135.944 internal mould	22.00	40.00	32.00
PMO 135.967 external mould	–	31.00	28.00
PMO 135.968 (holot.) int. mould	–	31.00	28.00
Ventral valve	s.l.	m.w.	h.w.
PMO 135.945 (parat.) int. mould	–	32.00	28.00

ridges crossing the muscle field. The extra-muscular areas are coarsely tuberculate and impressed by primary costae.

Interior of brachial valve: The cardinal-process lobes are ponderous and somewhat variably developed. Their attachment faces are high, thick, directed anterolaterally, and only slightly divergent. One specimen (the largest) has laterally curved faces; the others are straight. The bases converge towards and merge with a thick median ridge, creating a Y-shape. The degree of convergence seems to vary, so the arms of the Y are of variable length. (One specimen does not have bases that converge into the ridge). There is a deep pit or alveolus between and slightly anterior to the processes. One specimen shows a minuscule, thin ridge posteriorly between the processes. Thin, divergent socket ridges are present. They have 5–6 denticles on their posterior faces. Oblong, rounded muscle fields are deeply impressed posteriorly but feebly impressed anteriorly. They are delimited posteriorly by straight and thick ridges or calluses that encompass the width of the muscle field. There are a thin median ridge, or brevi-septum, and two variably developed, straight and nearly parallel trans-muscle ridges crossing and surpassing the muscle field. Another much fainter and shorter pair of ridges may branch out from the same posterior starting point, at a greater angle.

Remark. – The measurements for this large species are given in Table 48. The species is very rare, and there are few specimens in spite of very large bulk samples which were supplemented by extensive spot sampling.

Family Strophonellidae Hall, 1879

Genus *Strophonella* Hall, 1879

Subgenus *Eostrophonella* Williams, 1950

Type species. – *Amphistrophia davidsoni* Holtedahl, 1916. Lectotype selected by Williams (1951, p. 128); SM A30031; conjoined valves, the original of Davidson, 1871, Pl. 40:5, quoted by Holtedahl in his original description (1916, p. 64), from C1 Beds (Aeronian), bank of River Sefin, S of Llet-tyr'hyddod, Dyfed. Grid ref. SN 742 282.

Strophonella (Eostrophonella) davidsoni (Holtedahl, 1916)

Pls. 17:11–12, 14–15; 18:1–2, 4–5, 8, 11

Synonymy. – ☐1871 *Strophomena euglypha* Hisinger – Davidson, p. 288 (*pars*), Pl. 40:4, 5, *non* Pl. 40:1–3. ☐1916 *Amphistrophia davidsoni* sp. nov. – Holtedahl, p. 64 , Pl. 7:6–7. ☐1950 *Strophonella (Eostrophonella) davidsoni* (Holtedahl) – Williams, p. 282. ☐1951 *Strophonella (Eostrophonella) davidsoni* (Holtedahl) – Williams, p. 128, Pl. 8:15, 16. ☐1982 *Stropheodonta (Eostropheodonta) davidsoni* (Holtedahl) – Thomsen & Baarli, Pl. 2:14. ☐1986 *Strophonella (Eostrophonella) davidsoni* (Holtedahl) – Baarli, Fig. 6E.

Type material. – Same as genus.

Material. – PMO 103.524, 103.639, 105.884, 128.367, 128.368, 128.392, 130.317, 130.337, 130.346, 130.347, 130.352, 130.372: internal moulds of 2 pedicle and 6 brachial valves and 4 external moulds of valves from the Leangen Member, Solvik Formation, of Skytterveien and Leang-bukta, Asker.

Description. – Exterior: The valves are large, subtriangular and resupinate. Initially concavo-convex, they change to convexo-concave at a more mature stage. The width-to-length ratio is variable, but the material is consistently wider than long, with the majority only slightly wider. The maximum width is generally at the hinge line that tends to be extended as well-developed alae. The anterior margin also varies; it may be drawn out in a tongue or, in the wider specimens, merges evenly into the lateral margins. The anterior margin is thus rectimarginate or uniplicate. The beaks are obscure. The apsacline ventral interarea is relatively long, while the dorsal interarea is about $\frac{1}{2}$ as long and anacline. The delthyrium seems to be partly open. The chilidium is well developed and drawn down in a hood between the cardinal-process lobes.

The ornament is unequally parvicostellate, with 12 low, rounded primary costae per 10 mm at the 10 mm growth stage from the umbo. There are 2–5 secondary costellae between each costa at that growth stage.

Interior of pedicle valve: The ventral process is small, low and produced anteriorly into a thin myophragm. The dental plates are very short, triangular and stout. They have variably developed dental ridges that may curve anteriorly to delimit the muscle field. The dental plates diverge at about 20–30° to the hinge line. The muscle field is subcircular to oblong in outline and open anteriorly. The diductor scars are limited laterally by the dental ridges that may be slightly scalloped. The adductor scars are lanceolate and long, situated on each side of the thin myophragm. The external ornament may overprint the valve outside of the muscle area where coarse tuberculous are present.

Interior of brachial valve: The cardinalia is fairly delicate, compared with the size of the valve. The cardinal-process lobes are slim, with faces directed posterolaterally and partly covered by the chilidium. There is a long, very thin, and low ridge between them. The lobes diverge at 45° and are not joined proximally. The bases of the lobes diverge wider laterally. The sockets are narrow and shallow, bounded by long, thin socket ridges that diverge at about 20° to the hinge line. The sockets are highest at their anteroventral ends. A triangular notothyrial platform may be developed. If present, it continues as a low, broad median ridge that soon fades. Faintly impressed, elongate muscle scars may be seen posteriorly. Two very faint trans-muscle ridges may also be present, diverging at a slight angle to the median of the valve. The entire valve is overprinted by faint ribbing and is always covered by coarse tubercles.

Remarks. – Table 49 gives the measurements for this fairly large species. The negative covariation between sagittal length and maximum width in the dorsal valves probably reflects the great variation in shape found in the limited material.

Holtedahl (1916) erected the new species *Amphistrophia davidsoni* for this material. He selected, however, Davidson's (1871) specimens from Sefin Bridge, Dyfed, Wales, as type material. Holtedahl (1916) and later workers did not have pedicle valves from Norway. When Williams (1951) redescribed *S. (E.) davidsoni* from Wales, he concluded that Holtedahl's Norwegian material did not belong to that species and tentatively proposed inclusion in *S. (E.) eothen* Bancroft, 1949. The latter has recently been described both by Harper & Boucot (1978a) and Temple (1987). Clearly the Norwegian material does not belong to *S. (E.) eothen*. The pedicle valves collected in this study have much shorter dental plates and tend to have a more oblong ventral muscle field than that of *S. (E.) eothen*. The Norwegian pedicle valve agrees in these and other details well with that of *S. (E.) davidsoni*. The brachial valve does possess somewhat finer cardinal-process lobes and longer, finer socket ridges than *S. (E.) davidsoni*. However, these features are not as fine as those in *S. (E.) eothen*, and the socket ridges are widely

divergent as in *S. (E.) davidsoni*. The brachial valve thus has characteristics intermediate between *S. (E.) eothen* and *S. (E.) davidsoni*, but the general morphology is closer to the latter.

Williams (1951) particularly stressed the difference in age, the British material being of Late Llandovery age and the Norwegian of Early Llandovery age. However, the age difference is relatively minor. The material studied here was recovered from just below and in the lower parts of the *M. sedgwickii* graptolite zone. The exact level of Holtedahl's material from Malmøya is unknown. The type material is from the *M. sedgwickii* zone, which marks the earliest occurrence of the species in Wales. The ranges are therefore close or overlapping. The differences in morphology are so subtle that the material is best retained in *S. (E.) davidsoni*.

Superfamily Davidsoniacea King, 1850
Family Fardeniidae Williams, 1965
Genus *Saughina* Bancroft, 1949

Type species. – By original designation, *Schuchertella pertinax* Reed, 1917. Lectotype selected by Cocks (1978); B 72941; a pedicle valve, the original of Reed 1917, Pl. 19:23, from the Woodland Formation (Rhuddanian), Woodland Point, Girvan, Strathclyde. Grid ref. NX 168 953.

Saughina pertinax (Reed 1917) *gentilis* (Bancroft 1949)
Pl. 18:3, 6–7, 9–10, 12, 16

Type material. – Same as genus.

Synonymy. – ☐1917 *Schuchertella pertinax* – Reed (*pars*), p. 907, Pl. 19:23, Pl. 20:1, 2, 2a *non* Pl. 20:3. ☐1949 *Fardenia columbana gentilis* – Bancroft, p. 9, Pl. 2:13. ☐1949 *Saughina geoffreyi* – Bancroft, p. 7, Pl. 2:5. ☐1951 *Fardenia geoffreyi* [Bancroft MS] – Williams (*pars*), p. 120, Pl. 7:11–13, Text-figs. 11–13, 20a, b. ☐1970 *Fardenia* sp. – Temple, p. 51, Pl. 14:4–9, 12–13. ☐1978 *Fardenia (Saughina) geoffreyi* Bancroft, 1949 – Cocks, p. 134. ☐1978 *Fardenia (Saughina) pertinax* (Reed, 1917) – Cocks, p. 134. ☐1983 *?Fardenia* sp. – Lockley, p. 94, Figs. 2:9, 10. ☐1987 *Saughina pertinax* (Reed, 1917) morph. *gentilis* – Temple, pp. 90–92, Pl. 10:1–10.

Material. – PMO 103.472, 103.473, 103.513, 130.384, 130.385, 135.993: internal moulds of 1 pedicle and 4 brachial valves and 1 mould of valve exterior from the Myren Member at Spirodden and the Leangen Member, both Solvik Formation, at Skytterveien, Leangbukta, Asker and Jongsåsveien, Sandvika.

Description. – Exterior: Sub-circular, planoconvex valves with low total convexity, and ¾ as long as wide. The margins

Table 49. Strophonella (Eostrophonella) davidsoni (Holtedahl, 1916), measurements and statistics for valves.

Ventral valve	s.l.	m.w.	h.w.
PMO 103.639 internal mould	>24.00	39.00	39.00

Dorsal valves	s.l.	m.w.	h.w.
Mean	25.33	34.17	34.17
Std. dev.	4.89	4.40	4.40
Count	6	6	6
Minimum	18.00	28.00	28.00
Maximum	31.00	40.00	40.00
Variance–	23.87	0.50	0.50
covariance		19.37	19.37
matrix			19.37

are evenly rounded, and there are no pronounced alae. The maximum width occurs at or near the hinge line. The commissure is crenulated, and the anterior commissure is rectimarginate. Umbones are low, and the ventral beak is low and erect. A low, plane and triangular ventral interarea is apsacline. The dorsal interarea is very low in comparison, plane and anacline. There is a very small pseudodeltidium apically that continues in thin delthyrial plates. The delthyrial angle is about 60°. The notothyrial angles are about 80° and covered to $\frac{1}{3}$–$\frac{1}{2}$ by a small convex chilidium. The chilidium also partly covers the cardinal-process lobes. Both interareas are striated parallel to the hinge line.

Radial ornament is fine, parvicostellate, with 2–3 primary costae per millimeter at the 10 mm growth stage from umbo. There are 2–4 costellae between each costa. Very fine fila may be seen between the ribs.

Interior of pedicle valve: Small hinge teeth are present. They are supported by relatively strong and short dental plates. The dental plates diverge at about right angles, with the bases more strongly divergent than the tops. The inner faces of the dental plates have well-developed crural fossettes. The delthyrial chamber is shallow and delimited anteriorly by a break in the slope, behind which a pair of ex-sagittal ridges may represent the muscle scars. The scars are divided by a short, broad median ridge spanning the delthyrial chamber. There are no other signs of muscle scars, and most of the valve is overprinted by the external ornament.

Interior of brachial valve: The cardinalia are fairly stout, with cardinal-process lobes directed posteroventrally and projecting posterior of the hinge line. The lobes are oblong to rounded and partly covered by the convex chilidium. The lobes are separated posteriorly by a thin rounded groove which ends in an oblong process anteriorly. The sockets are small, fairly deep, and constricted by a small 'pad' of shell material on the hinge line and anteriorly by the high and long socket ridges. The socket ridges diverge anterolaterally at 80°. They are generally straight and splayed more at the bases than at the tops, but may curve slightly anterolaterally. The ridges are fused together anteromedially and descend steeply down onto the shallow notothyrial platform. The platform ends in a broad, low median ridge, which reaches forward to about $\frac{1}{4}$ of the valve length. Faint impressions of muscle scars may be visible between the median ridge and anterior points of the socket ridges.

Remarks. – The measurements for the present, fragmented and limited material are given in Table 50. The presence of a chilidium in this Norwegian material is crucial for generic placement. I follow Temple (1987) in considering *Chilidiopsis* Boucot (1959) a synonym of *Saughina*, awaiting further investigation of *Coolina* with respect to the presence of a chilidium. Thus, taxa with a well-developed chilidium, like that of the present material, are included in *Saughina*. (See under *Fardenia* for further comments.)

Table 50. *Saughina pertinax* (Reed, 1917) *gentilis* (Bancroft, 1949), measurements of valves.

Ventral valve	s.l.	m.w.	h.w.
PMO 103.472 internal mould	18.00	–	–

Dorsal valves	s.l.	m.w.	h.w.
PMO 103.473 internal mould	12.00	17.00	17.00
PMO 103.513 internal mould	19.00	–	–
PMO 130.384 internal mould	11.00	–	–
PMO 135.993 internal mould	17.00	–	–

This material is close to *Fardenia geoffreyi* Bancroft, 1949, as described by Williams (1951). Temple (1987) later redescribed the material and showed that *S. pertinax* (Reed, 1917) is the senior synonym. The Norwegian material clearly falls within the *S. pertinax gentilis* group, although it may differ in having a better differentiation of primary and secondary ribs than the Welsh material. Temple (1987), however, showed that the ribbing was subject to great variation.

Genus *Fardenia* Lamont, 1935

Type species. – By original designation *Fardenia scotica*. Lectotype selected by Cocks (1978); HML 1940, a pedicle valve, figured by Lamont, 1935, Pl. 7:1, from Lower Drummuck Group (Cautleyan), E brow of Quarrel Hill, Girvan, Stratclyde. Grid ref. NS 263 035.

Fardenia oblectator sp. nov.
Pl. 19:1–14

Type material. – Internal mould of a brachial valve, PMO 103.512 (Pl. 19:9) from the top of the Leangen Member, 3 m below the base of the Rytteråker Formation, Bleikerveien, Asker.

Derivation of name. – Latin *oblectator*, charmer, delighter.

Synonymy. – □1982 *Fardenia* sp. – Thomsen & Baarli, Pl. 2:9. □1987 *Fardenia* sp. – Baarli, Fig. 6d.

Material. – PMO 103.493 (3 specimens), 103.512, 109.725, 128.361, 128.373, 128.374, 130.370 (2 specimens), 130.373, 130.375–130.377, 130.379 (4 specimens), 130.380 (6 specimens), 130.381, 130.882, 130.883 (5 specimens), 130.885 (3 specimens), 130.887 (3 specimens), 130.888, 130.889 (4 specimens), 130.890: internal moulds of 15 pedicle and 25 brachial valves and external moulds of 3 valves from the Leangen Member, Solvik Formation, at Skytterveien, Bleikerveien, Leangbukta, Asker, Jongsåsveien, Sandvika and Bjerkøya, Holmestrand.

Diagnosis. – This is a small, biconvex *Fardenia* species with stout hinge teeth and relatively short dental plates with deep crural fossettes. The cardinal-process lobes are directed posteriorly and the socket ridges are relatively short and joined to a median ridge anteromedially.

Description. – Exterior: Relatively small, biconvex valves, the pedicle valve more convex than the brachial valve. The brachial valve is at least initially always convex but may flatten out anteriorly. The outline varies somewhat from subrectangular to subcircular and is between $3/5$ and $3/4$ as long as wide. The maximum width is at the hinge line, which may be extended laterally as short alae. The cardinal angles are acute and the lateral and anterior margins are variably rounded; the anterior margin has a tendency to be straighter than the sides. The anterior margin is rectimarginate, a few large valves being gently uniplicate because of the very gentle dorsal sulcus and ventral fold. The umbones are very low. The ventral interarea is straight, apsacline, and triangular, while the dorsal interarea is shorter and anacline. The delthyrium is open, with a tiny pseudodeltidium in a wide opening with 100–120° angle. The notothyrium seems to be open, except for the narrow chilidial plates.

The radial ornament is unequally parvicostellate, with 3 ribs per millimeter at the 10 mm growth stage from the umbo. The costellae arise by branching and intercalation producing a pattern where it may be difficult to distinguish between, in particular, costae and costella and where the ribs tend to bunch. The ribs are thin and rounded with flat interspaces.

Interior of pedicle valve: The teeth are large and triangular, and the posterior faces are flush with the posterior margin. They are supported ventrally by relatively short and laterally concave dental plates with deep crural fossettes. The bases of the dental plates are normally not drawn out anteriorly as ridges. Deep denticular cavities are formed between the dental plates and the hinge line. The delthyrial cavity is broad and gently concave. The muscle field is weakly impressed. Only the adductors are clearly observed. They are drawn out and pentagonal in outline, placed on each side of a weak median ridge. The diductor scars seem to be relatively small, oval and diverging laterally on each side of the adductor scars. Possibly they do not extend farther than the adductor scars. The muscle scars occupy $1/5$ of the total width and up to $2/5$ of the total length. The internal surface is strongly impressed by external ornament.

Interior of brachial valve: The cardinalia are raised high above the valve floor. The cardinal-process lobes are small, rounded, low, and directed posteriorly. They project slightly over the hinge line. The sockets are small, triangular and fairly deep. They are bounded anteriorly by high, straight, short and widely divergent (90–110°) socket ridges. These ridges are not joined anteromedially but are divided by a broad median ridge that descends from the front of the cardinal-process lobes and rapidly reaches the valve floor.

The median ridge may reach $1/4$ of the total valve length but is often shorter. The muscle fields are not observed, since the interior is strongly impressed by the external ornament.

Remarks. – Tables 51 and 52 show that *F. oblectator* is a relatively small species of *Fardenia*; the dorsal valves have a very high covariance for the sagittal length to maximum width, but the same covariance for the ventral valve is slightly low. Fig. 13 reveals that the ventral valves have a larger variation in shape for each growth stage than the dorsal valves. There are no indications of allometric growth in the dorsal valves.

This material may be distinguished from *Saughina pertinax* by the absence of a complete chilidium, less regular and coarser ornament, and cardinal-process lobes that are directed posteriorly as opposed to ventrally in the latter. The socket plates are also more divergent and do not join together anteromedially, as in *S. pertinax*.

This species is close to *Coolinia applanata* Salter, 1846, in ribbing, size and configuration of the socket ridges and cardinal-process lobes, as described by Williams (1951), Bassett (1974) and Cocks & Baarli (1982). Bassett (1974) claimed that *C. applanata* had a well-developed chilidium, seen also in Cocks & Baarli's (1982) Pl. 1:17. If a chilidium is present in the material described here, it is definitely smaller

Table 51. Fardenia oblectator sp. nov., measurements and statistics for the ventral valves.

	s.l.	m.w.	h.w.
PMO 103.493b (parat.) int. mould	10.00	18.00	18.00
Mean	11.00	16.42	16.42
Std. dev.	2.68	2.94	2.94
Count	11	12	12
Minimum	8.00	12.00	12.00
Maximum	15.00	21.00	21.00
Variance–	7.20	6.90	6.90
covariance		8.63	8.63
matrix			8.63

Table 52. Fardenia oblectator sp. nov., measurements and statistics for the dorsal valves.

	s.l.	m.w.	h.w.
PMO 103.493 (holot.) int. mould	9.00	15.00	15.00
Mean	11.23	16.03	15.97
Std. dev.	3.68	4.72	4.76
Std. error	0.95	1.08	1.09
Count	15	19	19
Minimum	4.50	7.50	7.50
Maximum	17.00	25.00	25.00
Variance–	13.53	19.52	19.61
covariance		22.29	22.46
matrix			22.68

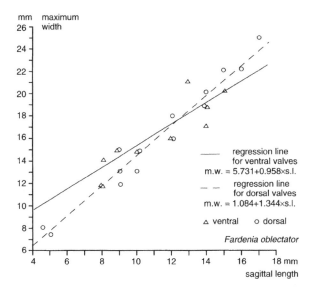

Fig. 13. Scattergram with regression lines for ventral and dorsal valves of *Fardenia oblectator* sp. nov., plotting sagittal length against maximum width.

than that figured by Cocks & Baarli (1982). Also, the present material has biconvex valves, as opposed to the resupinate pedicle and concave brachial valves of *C. applanata*. The delthyrial opening is also wider than in the latter. The present material is an older species than *C. applanata* and may be its ancestor. I tentatively place the Norwegian material in *Fardenia*, since I believe the chilidium is missing. I feel that the presence or absence of a chilidium ought not to be decisive for generic attribution, since this feature is often hard to observe. Also, as possibly in this case, it may obscure close generic relationships. If the presence of a chilidium is less important than previously thought, it might be that *Chilidiopsis*, *Fardenia* and *Saughina* are all synonyms of *Coolinia*. However, further work is needed with this group; for the present, I follow conventional wisdom.

Fardeniidae indet.

Pl. 18:13–15

Material. – PMO 130.344, 130.365, 130.367, 130.369: internal moulds of pedicle valves and 2 valve exteriors from the uppermost part of the Myren Member, Solvik Formation, at Sandvika, and the top of the Spirodden Member, Solvik Formation, at Spirodden.

Remarks. – Several fragmented valves of a fairly large and very thin, slightly concavo-convex fardiniid with fine parvicostellate ribbing. These seem to be most common in the Solvik Formation at Malmøya but are also found in the Spirodden Member, Solvik Formation, in Asker.

References

Amsden, T.W. 1951: Brachiopods of the Henryhouse Formation (Silurian) of Oklahoma. *Journal of Paleontology 25*, 69–96.

Amsden, T.W. 1968: Articulate brachiopods of the St. Clair Limestone (Silurian), Arkansas, and the Clarita Formation (Silurian), Oklahoma. *Journal of Paleontology, Memoir 42*, (part 2 of 2, supplement to No. 3), 1–117.

Amsden, T.W. 1974: Late Ordovician and early Silurian articulate brachiopods from Oklahoma, southwestern Illinois, and eastern Missouri. *Bulletin of the Oklahoma Geological Survey 119*, 1–154.

Baarli, B.G. 1985: The stratigraphy and sedimentology of the early Llandovery Solvik Formation, central Oslo Region, Norway. *Norsk Geologisk Tidsskrift 65*, 229–249.

Baarli, B.G. 1987: Benthic faunal associations in the Lower Silurian Solvik Formation, central Oslo Region, Norway. *Lethaia 20*, 75–90.

Baarli, B.G. 1988a: The Llandovery enteletacean brachiopods of the central Oslo region, Norway. *Palaeontology 31*, 1101–1129.

Baarli, B.G. 1988b: Bathymetric coordination of proximality trends and level-bottom communities: a case study from the Lower Silurian of Norway. *Palaios 3*, 577–587.

Baarli, B.G. 1990a: Depositional environments in the Telychian Stage (Silurian) of the central Oslo region, Norway. *Geological Journal 25*, 65–79.

Baarli, B.G. 1990b: Peripheral bulge of a foreland basin in the Oslo Region during the early Silurian. *Palaeogeography, Palaeoclimatology, Palaeoecology 78*, 149–161.

Baarli, B.G., Brande, S. & Johnson, M.E. 1992: Proximality trends in the Red Mountain Formation (Lower Silurian) of Birmingham, Alabama. *Oklahoma Geological Survey, Bulletin 145*, 1–13.

Baarli, B.G. & Harper, D.A.T. 1986: Relict Ordovician brachiopod faunas in the Lower Silurian of Asker, Oslo Region, Norway. *Norsk Geologisk Tidsskrift 66*, 87–98.

Baarli, B.G. & Johnson, M.E. 1988: Biostratigraphy of selected brachiopods from the Llandovery Series (Lower Silurian) of the Oslo Region. *Norsk Geologisk Tidsskrift 67*, 259–274.

Bancroft, B.B. 1949: *Welsh Valentian Brachiopods and the* Strophomena antiquata *Group of Fossil Brachiopods.* 16 pp. Privately printed (edited by A. Lamont). Mexborough.

Barrande, J. 1879: *Systeme silurien du Centre de la Boheme,* V, 226pp, Pls.153, Praha and Paris.

Bassett, M.G. 1970: The articulate brachiopods from the Wenlock Series of the Welsh Borderland and South Wales. Part 1. *Palaeontographical Society (London), Monograph 123 (525)*, 1–26.

Bassett, M.G. 1972: The articulate brachiopods from the Wenlock Series of the Welsh Borderland and South Wales. Part 2. *Palaeontographical Society (London), Monograph 126 (532)*, 27–78.

Bassett, M.G. 1974: The articulate brachiopods from the Wenlock Series of the Welsh Borderland and South Wales. Part 3. *Palaeontographical Society (London), Monograph 128 (541)*, 79–123.

Bassett, M.G. & Cocks, L.R.M. 1974: A review of Silurian brachiopods from Gotland. *Fossils and Strata 3*, 1–56.

Bergström, J. 1968: Upper Ordovician brachiopods from Västergötland, Sweden. *Geologica et Palaeontologica 2*, 1–35.

Billings, E. 1861–1865: Palaeozoic Fossils: containing descriptions and figures of new or little known species of organic remains from the Silurian rocks. *Canadian Geological Survey 1*. 426 pp.

Boucot, A.J. 1959: A new family and genus of Silurian orthotetacid brachiopods. *Journal of Paleontology 33*, 25–28.

Boucot, A.J. 1960: Lower Gedinnian brachiopods of Belgium. *Louvain University, Instute of Geology, Memoir 21*, 281–324.

Boucot, A.J., Johnson, J.G., Harper, C.W. & Walmsley, V.G. 1966: Silurian brachiopods and gastropods of southern New Brunswick. *Bulletin of the Geological Survey of Canada 140*, 1–45.

Brenchley, P.J. & Cocks, L.R.M. 1982: Ecological associations in a regressive sequence: the latest Ordovician of the Oslo–Asker district – Norway. *Palaeontology 25*, 783–815.

Caster, K.E. 1939: A Devonian fauna from Colombia. *Bulletin of American Paleontology 24:83*. 218 pp.

Chiang, K.K. 1972: *Hesperorthis* and two new Silurian species. *Journal of Paleontology 46:3*, 353–359.

Cocks, L.R.M. 1967: Llandovery Stropheodontids from the Welsh Borderland. *Palaeontology 10:2*, 245–265.

Cocks, L.R.M. 1968: Some Strophomenacean brachiopods from the British Lower Silurian. *British Museum (Natural History), Bulletin, (Geology) 15:6*, 285–324.

Cocks, L.R.M. 1970: Silurian brachiopods of the superfamily Plectambonitacea. *British Museum (Natural History), Bulletin (Geology) 19:4*, 142–203.

Cocks, L.R.M. 1978: A review of British Lower Palaeozoic brachiopods, including a synoptic revision of Davidson's monograph. *Palaeontographical Society (London), Monograph 131 (549)*. 256 pp.

Cocks, L.R.M. 1982: The commoner brachiopods of the latest Ordovician of the Oslo–Asker District, Norway. *Palaeontology 25*, 755–781.

Cocks, L.R.M. & Baarli, B.G. 1982: The Late Llandovery brachiopods from the Oslo Region. *In* Worsley, D. (ed.): Field meeting, Oslo Region 1982. *Paleontological Contributions from the University of Oslo 278*, 79–89.

Cocks, L.R.M. & Fortey, R.A. 1982: Faunal evidence for oceanic separations in the Paleozoic in Britain. *Journal of the Geological Society of London 139*, 465–478.

Cocks, L.R.M. & Rong, J.-y. 1989: Classification and review of the brachiopod superfamily Plectambonitacea. *British Museum (Natural History), Bulletin (Geology) 45:1*, 77–163.

Cocks, L.R.M., Woodcock, N.H., Rickards, R.B., Temple, J.T. & Lane, P.D. 1984: The Llandovery Series of the type area. *British Museum (Natural History), Bulletin (Geology) 38:3*, 131–182.

Conrad, T.A. 1843: Observations on the lead bearing limestone of Wisconsin and descriptions of a new genus of trilobite and descriptions of organic remains. *The Philadelphia Academic Journal of Natural Sciences 1842*, 329–335.

Cooper, G.A. 1956: Chazyan and related brachiopods. *Smithsonian Miscellaneous Collections, Washington 127*. 1245 pp.

Dalman, J.W. 1828: Uppställning och beskrifning af de i Sverige funne Terebratuliter. *Kungliga Svenska Vetenskaps-Akademiens Handlingar 3*, 85–155.

Davidson, T. 1848: Memoire sur les brachiopodes du système superiur d'Angleterre. *Bulletin de la Societe géologique de France (2) 5*, 309–338.

Davidson, T. 1869: A monograph of the British fossil Brachiopoda. Part 7 (3), The Silurian Brachiopoda. *Palaeontographical Society, (Monograph)*, 169–248.

Davidson, T. 1871: A monograph of the British fossil Brachiopoda. Part 7 (4). The Silurian Brachiopoda. *Palaeontographical Society (Monograph)* 249–397.

Davidson, T. 1883: Monograph of the British fossil Brachiopoda. Vol 5, Part 2. Silurian supplements. *Palaeontographical Society (Monograph)*, 135–242.

Foerste, A.F. 1909: Fossils from the Silurian formations of Tennessee, Indiana and Illinois. *Denison University, Scientific laboratories, Bulletin 14*, 61–107.

Foerste, A.F. 1912: *Strophomena* and other fossils from Cincinnatian and Mohawkian horizons, chiefly in Ohio, Indiana, and Kentucky. *Journal of Paleontology 44*, 125–132.

Fortey, R.A. & Cocks, L.R.M. 1992: The early Palaeozoic of the North Atlantic regions as a test case for the use of fossils in continental reconstruction. *Tectonophysics 206*, 147–158.

Hall, J. 1843: Geology of New York. Part 4, comprising the survey of the Fourth Geological District. *Natural History of New York 4*, 1–683.

Hall, J. 1847: Description of the organic remains of the lower division the New York System. *New York State Geological Survey, Paleontology, New York 1, 23*. 338 pp.

Hall, J. 1857: Descriptions of Paleozoic fossils. *Annual report of New York State Cabinet of Natural History 10*, 39–180.

Hall, J. 1879: The fauna of the Niagara Group. *New York State Museum of Natural History, 28th report*, 93–203.

Hall, J. 1883: Brachiopoda, plates and explanations. *Annual Report of the New York States Geologist 4:2*, Pls. 34–61.

Hall, J. & Clarke, J.M. 1892: An introduction to the study of the genera of Paleozoic Brachiopoda. Part 1. *New York State Geological Survey, Paleontology of New York 8*. 367 pp.

Hall, J. & Clarke, J.M. 1894: An introduction to the study of the Brachiopoda. *12th Annual report of the New York State Geologist, Palaeontology 2*, 751–943.

Harper, C.W. 1973: Brachiopods of the Arisaig Group (Silurian–Lower Devonian) of Nova Scotia. *Bulletin of the Geological Society of Canada 215*, 1–163.

Harper, C.W. & Boucot, A.T. 1978a: The Stropheodontacea, Part 1 (Leptostrophiidae, Eostropheodontidae and Strophonellidae). *Palaeontographica A 161:1–3*, 55–118.

Harper, C.W. & Boucot, A.J. 1978b: The Stropheodontacea, Part 2 (Douvillinidae, Telaeoshaleriidae, Amphistrophiidae and Shaeriidae). *Palaeontographica A 161:4–6*, 119–175.

Harper, C.W. & Boucot, A.J. 1978c: The Strophodontacea, Part 3 (Strophodontidae (*sensu strictu*), Pholidostrophiidae and Lissostrophiidae). *Palaeontographica A 162:1–2*, 1–80.

Harper, C.W., Johnson, J.G. & Boucot, A.J. 1967: The Pholidostrophiinae (Brachiopoda: Ordovician, Silurian, Devonian). *Senkenbergiana Lethaea 48*, 403–461.

Harper, D.A.T. 1984: Brachiopods from the Upper Ardmillian succession (Ordovician) of the Girvan district, Scotland. Part 1. *Palaeontographical Society (London) Monograph 136 (565)*. 78 pp.

Havlíček, V. 1961: Plectambonitacea im Böhmischen Paläozoikum (Brachiopoda). *Věstnik Ústředniho Ústavu Geologického 36:6*, 447–451.

Havlíček, V. 1977: Brachiopods of the order Orthidae in Czechoslovakia. *Ústřední Ústav Geologický, Rozpravy 44*, 1–327.

Havlíček, V. & Štorch, P. 1990: Silurian brachiopods and benthic communities in the Prague Basin (Czechoslovakia). *Ústřední Ústav Geologický, Rozpravy 48*. 275 pp.

Holtedahl, O. 1916: Strophomenidae of the Kristiania Region. *Videnskabsselskapets Skrifter, I, Mathematisk-naturvidenskabelige Klasse 1915. Kristiania 12*. 118 pp.

Holtedahl, O. & Dons, J.A. 1952: Geologisk kart over Oslo og omegn. *Supplement to: Geological Guide to Oslo and District.* Universitetsforlaget, Oslo.

Hurst, J.M. 1974: Aspects of the systematics and ecology of the brachiopod *Pholidostrophia* in the Ashgill, Llandovery and Wenlock of Britain. *Neues Jahrbuch fuer Geologie und Palaeontologie Abhandlungen 146*, 298–324.

Huxley, T.H. 1869: *An Introduction to the Classification of Animals.* 147 pp. London.

Jones, O.T. 1928: *Plectambonites* and some allied genera. *Memoirs of the Geological Survey of Great Britain, Palaeontology 1:5*, 367–527.

Kiaer, J. 1908: Das Obersilur im Kristianiagebiete. Eine stratigraphisch-faunistische Untersuchung. *Skrifter fra Videnskabs-Selskabet i Christiania 1906, I Mathematisk-Naturvidenskaplige Klasse 2*. 595 pp.

King, W. 1846: Remarks on certain genera belonging to the class Palliobranchiata. *Annals and Magazine of Natural History, Series 1, 18*, 26–42, 83–94.

King, W. 1850: A monograph of the Permian fossils of England. *Palaeontographical Society (London), Monograph 3*. 258 pp.

Koren, T.N., Oradovskaya, M.M. & Pylma, L.Y., Sobolevskaya, R.F. & Chugaeva, M.N. 1983: *The Ordovician–Silurian Boundary in the Northeast of the USSR.* 192 pp. Nauka, Leningrad.

Kozłowski, R. 1929: Les brachiopodes gothlandiens de la Podolie Polonaise. *Palaeontologia Polonica 1*. 254 pp.

Kul'kov, N.P. & Severgina, L.G. 1989: Stratigrafiya i brakhiopody Ordoviki i nizhnego Silura Gornogo Altaya. [Stratigraphy and brachiopods of the Ordovician and Lower Silurian in Gornyj Altaj.] *Akademiia Nauk SSSR, Sibirskoe Otdelenie, Institut Geologii i Geofiziki (IGIG) Trudy 717*. 221.

Lamont, A. 1935: The Drummuck Group: a stratigraphical revision, with descriptions of new fossils from the lower part of the Group. *Transactions of the Geological Society of Glasgow 19*, 288–332.

Lamont, A. & Gilbert, D.L.F. 1945: Upper Llandovery Brachiopoda from Coneygore Coppice and Old Storridge Common, near Alfrick, Worcestershire. *Annals and Magazine of Natural History, Series 11, 12:94*, 641–682.

Lindström, G. 1861: Bidrag till kännedomen om Gotlands brachiopoder. *Öfversigt af Kungliga Vetenskaps-Akademiens Förhandlingar 17 (for 1860)*, 337–382.

Lockley, M.G. 1983: Brachiopods from a lower Palaeozoic mass flow deposit near Llanafan Fawr, central Wales. *Geological Journal 18:1*, 93–99.

McCoy, F. 1846: *A Synopsis of the Silurian Fossils of Ireland Collected from the Several Districts by Richard Griffith, F.G.S.* 72 pp. Dublin.

McEwan, E.D. 1919: A study of the brachiopod genus *Platystrophia*. *United States National Museum, Proceedings 56 (2297)*, 383–448.

Möller, N.K. 1989: Facies analysis and paleogeography of the Rytteråker Formation (Lower Silurian; Oslo Region, Norway). *Palaeogeography, Palaeoclimatology, Palaeoecology 69*, 167–192.

Öpik, A. 1930: Brachiopoda Protremata der Estlandischen Ordovischen Kukruse-Stufe. *Tartu Universitatis (Dorpatensis), Acta et Commentationes, Series A, 17:1.* 262 pp.

Öpik, A. 1933: Über Plectamboniten. *Tartu Universitatis (Dorpatensis), Acta et Commentationes, Series A, 24:7.* 79 pp.

Öpik, A. 1934: Über Klitamboniten. *Tartu Universitatis (Dorpatensis), Acta et Commentationes, Series A, 26:3.* 239 pp.

Paeckelman, W. & Sieverts, H. 1932: Neue Beiträge zur Kenntnis der Geologie, Palaeontologie, und Petrographie der Umgegend von Konstantinopel. 1. Obersilurische und Devonische Faunen der Prinzeninseln, Bithyniens und Thraziens. *Königliche–Preussiche Geologische Landesanstalt, Abhandlungen 142.* 79 pp.

Percival, I.G. 1991: Late Ordovician Articulate brachiopods from central New South Wales. *In* Jell, P.A. (ed.): Australian Ordovician brachiopod studies. *Memoir of the Association of Australasian Palaeontologists 11*, 107–177.

Phillips, J. & Salter, J.W. 1848: Palaeontological appendix to Professor John Phillips' Memoir on the Malvern Hills compared with the Palaeozoic districts of Abberley, etc. *Memoirs of the Geological Survey of Great Britain 2:1*, 331–386.

Popov, L.E., Bassett, M.G., Holmer, L.E. & Laurie, J. 1993: Phylogenetic analysis of higher taxa of Brachiopoda. *Lethaia 26*, 1–6.

Potter, A.W. 1990: Middle and Late Ordovician brachiopods from the eastern Klamath Mountains, Northern California. Part 1 *Palaeontographica A 212*, 31–158.

Reed, F.R.C. 1917: The Ordovician and Silurian Brachiopoda of the Girvan district. *Royal Society of Edinburgh, Transactions 51:4*, 795–998.

Rong, J.-y. & Cocks, L. R. M. 1994: True *Strophomena* and a revision of the classification and evolution of strophomenoid and 'strophodontoid' brachiopods. *Palaeontology 37*, 651–694.

Rong, J-y. & Harper, D.A.T. 1988: A global synthesis of the latest Ordovician Hirnantian brachiopod fauna. *Transactions of the Royal Society of Edinburgh: Earth Science 79*, 383–402.

Ross, R.J. 1959: Brachiopod fauna of Saturday Mountain Formation, Southern, Lemhi Range, Idaho. *Professional Papers from the Geological Survey 294-L*, 441–461. Washington, D.C.

Rubel, M. 1962: Brakhiopody Orthacea Llandoveri Estonii. (Estonian Llandoverian brachiopods of the superfamily Orthacea.) *Eesti NSV teaduste Akadeemia Geologia instituudi, uurimused 9*, 75–94, Pls. 1–4.

Rubel, M. 1963: Brakhiopody Orthida Silura Estonii. *Eesti NSV Teaduste Akadeemia Geologia Instituudi, Uurimused 13*, 109–160.

Salter, J.W. 1846. Addenda. *In* McCoy, F.: *A Synopsis of the Silurian Fossils of Ireland Collected from the Several Districts by Richard Griffith, F.G.S.*, 69–79. Dublin.

Schlotheim, E.F. von 1820: *Die Petrefactenkunde auf ihrem jetzigen Standpunkte durch die Beschreibungen einer Sammlung versteinerter und fossiler Uberreste der Tier- und Pflanzenreichs der Vorwelt erläutert 1.* lxii + 378 pp. Gotha.

Schmidt, F. 1858: Untersuchungen Über die silurische Formation von Ehstland, Nord Livland und Oesel. *Archiv für Naturkunde Livland, Ehstland und Kurlands (Estonian SSR), Series I, Bd.II.*

Schuchert, C. 1893: Classification of the Brachiopoda. *American Geologist 11*, 141–167.

Schuchert, C. & Cooper, G.A. 1931: Synopsis of the brachiopod genera of the sub-orders Orthoidea and Pentameroidea, with notes on the Telotremata. *American Journal of Science, Series 5, 22*, 241–251.

Schuchert, C. & Cooper, G.A. 1932: Brachiopod genera of the suborders Orthoidea and Pentameroidea. *Peabody Museum of Natural History 4:1.* 270 pp.

Schuchert, C. & Le Vene, C.M. 1929: *Fossilum Catalogus, 1, Animalia. Pars 42. Brachiopoda (Generum et genotyporum index et bibliographia).* 140 pp. Junk, Berlin.

Şengör, A.M.C., Natal'in, B.A. & Burtman, V.S. 1993: Evolution of the Altaid tectonic collage and Palaeozoic crustal growth in Eurasia. *Nature 364*, 299–307.

Shaler, N.S. 1865: List of the Brachiopoda from the island of Anticosti, sent by the Museum of Comparative Zoology to different institutions in exchange for other specimens, with annotations. *Harvard University, Museum of Comparative Zoology, Bulletin 1*, 61–70.

Sowerby, J. de C. 1839: [Description of fossil shells.] *In* Murchison, R.I.: *The Silurian System, Founded on Geological Researches in the Counties of Salop, Hereford, Radnor, Montgomery, Caermarthen, Brecon, Pembroke, Monmouth, Gloucester, Worcester, and Stafford; with Descriptions of the Coalfields and Overlying Formations. Part 2.* 786 pp. London.

Spjeldnaes, N. 1957: The Middle Ordovician of the Oslo Region, Norway. 8. Brachiopods of the Suborder Strophomenida. *Norsk Geologisk Tidsskrift 37*, 1–214.

Stainbrook, M.A. 1943: Strophomenacea of the Cedar Valley limestone of Iowa. *Journal of Paleontology 17*, 39–59.

Teichert, C. 1928: Stratigraphische und paläontologische Untersuchungen im unteren Gotlandium (Tamal-Stufe) des westlichen Estland und der Insel Dagö. *Neues Jahrbuch für Mineralogie und Geologie 60(B).* 112 pp.

Temple, J.T. 1968: The Lower Llandovery (Silurian) brachiopods from Keisley, Westmorland. *Palaeontological Society (London), Monograph 122 (521).* 58 pp.

Temple, J.T. 1970: The Lower Llandovery brachiopods and trilobites from Ffridd Matrafal, near Montgomeryshire. *Palaeontographical Society (London), Monograph 124 (527).* 76 pp.

Temple, J.T. 1987: Early Llandovery brachiopods of Wales. *Palaeontographical Society (London), Monograph 139 (572).* 137 pp.

Thomsen, E. & Baarli, B.G. 1982: Brachiopods of the Lower Llandovery Saelabonn and Solvik formations of the Ringerike, Asker, and Oslo districts. *In* Worsley, D. (ed.): *Field Meeting, Oslo Region 1982*, 63–78. *Paleontological Contributions of the University of Oslo 278.*

Torsvik, T.H., Olesen, O., Ryan, P.D. & Trench, A. 1990: On the palaeogeography of Baltica during the Palaeozoic: new palaeomagnetic data from the Scandinavian Caledonides. *Geophysical Journal International 103*, 261–279.

Ulrich, E.O. & Cooper, G.A. 1942: New genera of Ordovician brachiopods. *Journal of Paleontology 16*, 620–626.

Wahlenberg, G. 1818: Petrificata Telluris Suecanae examinata. *Nova Acta Regiae Societatis Scietiarum Upsaliensis (for 1821) 8.* 116 pp.

Walmsley, V.G. & Boucot, A.J. 1975: The phylogeny, taxonomy and biogeography of Silurian and early to mid Devonian Isorthinae (Brachiopoda). *Palaeontographica A 148*, 34–108.

Whittard, W.F. & Barker, G.H. 1950: The Upper Valentian brachiopod fauna of Shropshire. I. Inarticulata: Articulata, Protremata, Orthoidea. *Annals and Magazine of Natural History, Series 12, 3:31*, 553–590.

Williams, A. 1950: New stropheodontid brachiopods. *Washington Academy of Science, Journal 40*, 277–282.

Williams, A. 1951: Llandovery brachiopods from Wales with special reference to the Llandovery District. *Geological Society of London, Quarterly Journal 107:1*, 85–136.

Williams, A. 1965: Subfamily Furcitellinae. *In* Moore, R.C. (ed.): *Treatise on Invertebrate Paleontology, Part H, Brachiopoda 1*, H384–386. Geological Society of America, Boulder, Col., and University of Kansas Press, Lawrence, Ka.

Williams, A. & Wright, A.D. 1965: The Orthida. *In* Moore, R.C. (ed.): *Treatise on Invertebrate Paleontology, Part H, Brachiopoda 1*, H299–H359. Geological Society of America, Boulder, Col., and University of Kansas Press, Lawrence, Ka.

Williams, A. & Wright, A.D. 1981: The Ordovician–Silurian boundary in the Garth area of southwest Powys, Wales. *Geological Journal 16*, 1–39.

Witzke, B.J. 1990: Palaeoclimatic constraints for Palaeozoic Palaeolatitudes of Laurentia and Euramerica. *In* McKerrow, W.S. & Scotese, C.T. (eds.): *Palaeozoic Palaeogeography and Biogeography. Geological Society Memoir 12*, 57–73.

Wright, A.D. 1968: The brachiopod *Dicoelosia biloba* (Linnaeus) and related species. *Arkiv för Zoologi 20*, 261–319.

Woodward, S.P. 1852: *A Manual of Mollusca 2*, 159–330. London.

Plate 1

1–3, 5–7, 9–11, 14, 15. ORTHOKOPIS IDUNNAE sp. nov. pp. 9–10.

☐1, 2. PMO 128.153: Holotype. Nine meters above the base of the Myren Member, Solvik Formation, Konglungø, Asker; ×3. Internal mould (1) and latex cast (2) of brachial valve.

☐3, 7. PMO 128.104: Fifty meters above the base of the Myren Member, Solvik Formation, Spirodden, Asker; ×4. Internal mould (3) and latex cast (7) of pedicle valve.

☐5. PMO 128.096. At the base of the Myren Member, Solvik Formation, Nesøya, Asker; ×3. Internal mould of pedicle valve.

☐6, 10. PMO 128.199. At the base of the Myren Member, Solvik Formation, Nesøya, Asker; ×3. Internal mould (6) and latex cast (10) of brachial valve.

☐9. PMO 128.201. At the base of the Myren Member, Solvik Formation, Nesøya, Asker; ×4. Internal mould of pedicle valve.

☐11, 14, 15. PMO 128.198. Three meters above the base of the Myren Member, Solvik Formation, Konglungø, Asker; ×3. Dorsal (11), posterior (14), and ventral (15) view of whole valve.

4, 8, 12–13, 16, 19, 22, 25. DOLERORTHIS SOWERBYIANA (Davidson, 1869), pp. 10–11.

☐4. PMO 103.516. Ten meters above the base of the Myren Member, Solvik Formation, Spirodden, Asker; ×2. Internal mould of pedicle valve.

☐8. PMO 128.204. At the base of the Myren Member, Solvik Formation, Hvalsodden, Asker; ×2. External view of pedicle valve.

☐12. PMO 128.106. Nine meters above the base of the Myren Member, Solvik Formation, Konglungø, Asker; ×2.5. Internal mould of pedicle valve.

☐13. PMO 128.238. At the base of the Myren Member, Solvik Formation, Nesøya, Asker; ×3. Internal mould of brachial valve.

☐16, 19. PMO 128.114. Nine meters above the base of the Myren Member, Solvik Formation, Konglungø, Asker; ×3. Latex cast (16) and internal mould (19) of brachial valve.

☐22. PMO 128.113. Nine meters above the base of the Myren Member, Solvik Formation, Konglungø, Asker; ×3. Internal mould of brachial valve.

☐25. PMO 128.117. Six meters above the base of the Myren Member, Solvik Formation, Vakås, Asker; ×3. External mould of pedicle valve.

17–18, 20–21, 23–24. DOLERORTHIS aff. SOWERBYIANA (Davidson, 1869), pp. 11–12.

☐17. PMO 128.139. At the base of the Leangen Member, Solvik Formation, Skytterveien, Asker; ×2. Internal mould of brachial valve.

☐18. PMO 128.142. Four meters above the base of the Leangen Member, Solvik Formation, Skytterveien, Asker. Internal mould of brachial valve.

☐20–21. PMO 128.138. At the base of the Leangen Member, Solvik Formation, Skytterveien, Asker; ×2. External mould (20) and latex cast (21) of brachial valve.

☐23, 24. PMO 128.140. At the base of the Leangen Member, Solvik Formation, Skytterveien, Asker; ×2. Latex cast (23) and internal mould (24) of pedicle valve.

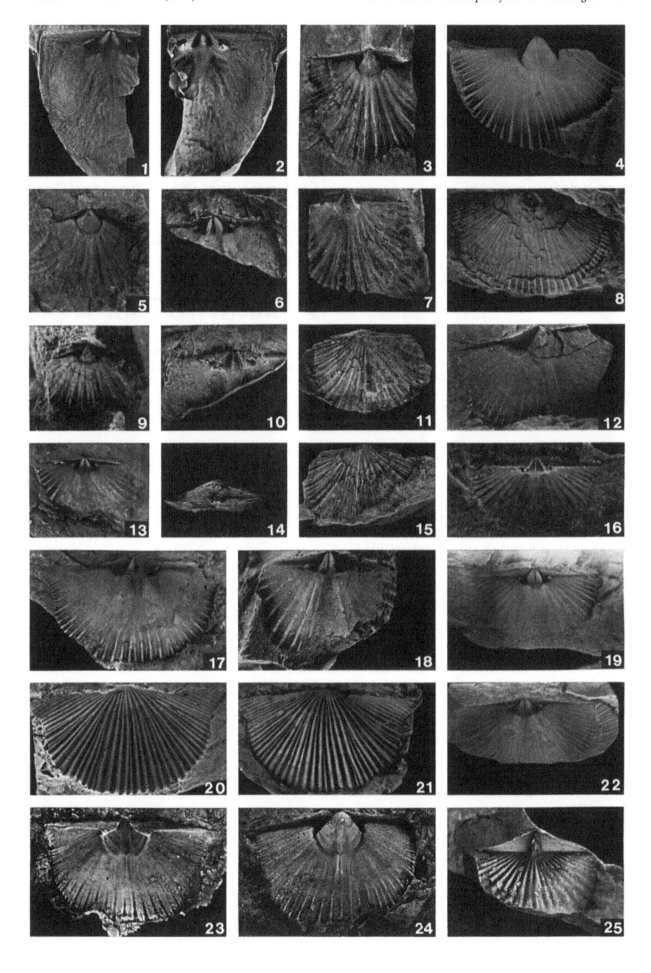

Plate 2

1–2, 4–5. DOLERORTHIS aff. SOWERBYIANA (Davidson, 1869), pp. 11–12.

□1. PMO 128.160. Six meters above the base of the Leangen Member, Solvik Formation, Skytterveien, Asker; ×2. Internal mould of gerontic pedicle valve.

□2. PMO 128.142. Four meters above the base of the Leangen Member, Skytterveien, Asker; ×2.5. Latex cast of brachial valve.

□4, 5. PMO 128.145. At the base of the Leangen Member, Solvik Formation, Asker; ×2. Internal mould (4) and latex cast (5) of brachial valve.

3, 6, 9, 12, 15–18. SCHIZONEMA SUBPLICATUM (Reed, 1917), pp. 13–14.

□3, 6. PMO 128.098. Seventeen meters above the base of the Myren Member, Solvik Formation, Drammensveien, Vakås, Asker; ×2. Internal mould (3) and latex cast (6) of brachial valve.

□9, 12. PMO 128.094. At the base of the Myren Member, Solvik Formation, Nesøya, Asker; ×2. Latex cast (9) and external mould (12) of brachial valve.

□15, 18. PMO 107.592. Seventeen meters above the base of the Myren Member, Solvik Formation, Drammensveien, Vakås, Asker; ×2.5. Latex cast (15) and internal mould (18) of brachial valve. Refigured from Baarli & Harper (1986, Pl. 1d, e).

□16, 17. PMO 128.095. Base of the Myren Member, Solvik Formation, Nesøya, Asker; ×2.5. Latex cast (16) and internal mould (17) of pedicle valve.

7–8, 10–11, 13–14. DOLERORTHIS aff. PSYGMA Lamont & Gilbert, 1945, pp. 12–13.

□7, 11. PMO 43.391. Uppermost part of the Rytteråker Formation, about 30 m up in the profile at the point west of Skinnerbukta, Malmøya, Oslo; ×1.5. Internal mould (7) and latex cast (11) of brachial valve.

□8. PMO 139.136. Counterpart to PMO43.389. Uppermost part of the Rytteråker Formation, about 30 m up in the profile at the point west of Skinnerbukta, Malmøya, Oslo; ×1.5. External mould of brachial valve.

□10, 13. PMO 43.445. Lowermost parts of the Vik Formation, Malmøya, Oslo; ×1.5. Latex cast (10) and internal mould (13) of pedicle valve.

□14. PMO 43.447. Lowermost parts of the Vik Formation, Malmøya, Oslo: ×1.5. External pedicle valve.

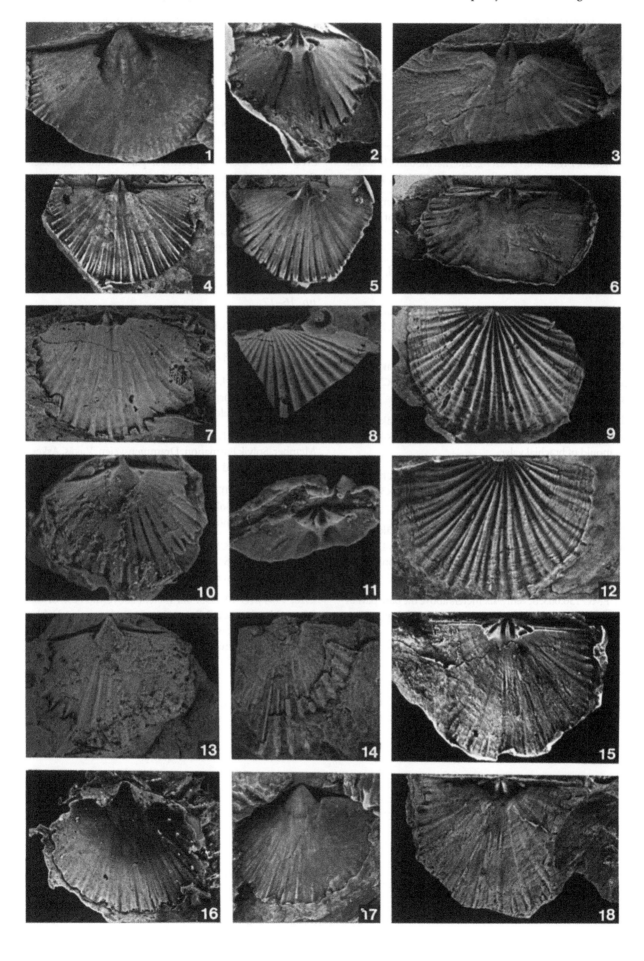

Plate 3

1–2. Schizonema subplicatum (Reed, 1917),
pp. 13–14.

☐1. PMO 105.200. At the base of the Myren Member, Solvik Formation, Brønnøya, Asker; ×2. Internal mould of brachial valve. Refigured from Thomsen & Baarli (1982, Pl. 1:7).

☐2. PMO 105.202. At the base of the Myren Member, Solvik Formation, Brønnøya, Asker; ×2. Internal mould of pedicle valve.

3–10, 12–13. Hesperorthis hillistensis Rubel, 1962,
p. 15.

☐3. PMO 109.889. Uppermost parts of the Rytteråker Formation, Christian Skredsviks vei, Valler Bærum; ×2. Internal mould of pedicle valve.

☐4. PMO 135.985. Uppermost parts of the Rytteråker Formation, Christian Skredsviks vei, Valler, Bærum; ×2. Internal mould of pedicle valve, lateral view.

☐5. PMO 135.954. Uppermost parts of the Rytteråker Formation, Christian Skredsviks vei, Valler, Bærum; ×4. Internal mould of brachial valve.

☐6. PMO 135.952. Uppermost parts of the Rytteråker Formation, Christian Skredsviks vei, Valler, Bærum; ×3. Internal mould of brachial valve.

☐7. PMO 135.961. Uppermost parts of the Rytteråker Formation, Christian Skredsviks vei, Valler, Bærum; ×4. External mould of brachial valve. Mark the fine striations of the costa.

☐8. PMO 136.005. Seventy-six meters above the base part of the Rytteråker Formation, Kampebråten, Sandvika; ×3. External mould of brachial valve.

☐9–10. PMO 135.989. Uppermost parts of the Rytteråker Formation, Christian Skredsviks vei. Valler, Bærum; ×6. Internal mould (9) and latex cast (10) of brachial valve.

☐12. PMO 135.950. Uppermost parts of the Rytteråker Formation, Christian Skredsviks vei, Valler, Bærum; ×2. Internal mould of pedicle valve.

☐13. PMO 135.990. Uppermost parts of the Rytteråker Formation, Christian Skredsviks vei, Valler Bærum; ×4. Latex cast of pedicle valve.

11, 14–20 ?Hesperorthis gwalia (Bancroft, 1949),
pp. 16–17.

☐11. PMO 130.923. One meter above the base of the Myren Member, Solvik Formation, Drammensveien, Vakås, Asker; ×2. Internal mould of pedicle valve.

☐14. PMO 130.894. Six meters above the base of the Myren Member, Solvik Formation, Drammensveien, Vakås, Asker; ×2.5. Internal mould of pedicle valve.

☐15. PMO 128.121. Six meters above the base of the Myren Member, Solvik Formation, Drammensveien, Vakås, Asker; ×3. Internal mould of brachial valve.

☐16. PMO 128.107. Nine meters above the base of the Myren member, Solvik Formation, Konglungø, Asker; ×3. Latex of internal mould of pedicle valve.

☐17, 20. PMO 128.216. Fifteen meters above the base of the Myren Member, Solvik Formation, Spirodden, Asker; ×2. Latex cast (17) and internal mould (20) of brachial valve.

☐18, 19. PMO 107.595. Six meters above the base of the Myren Member, Solvik Formation, Drammensveien, Vakås, Asker; ×3. Latex cast (18) and external mould (19) of brachial valve. Refigured from Baarli & Harper (1986, Pl. 1h, l).

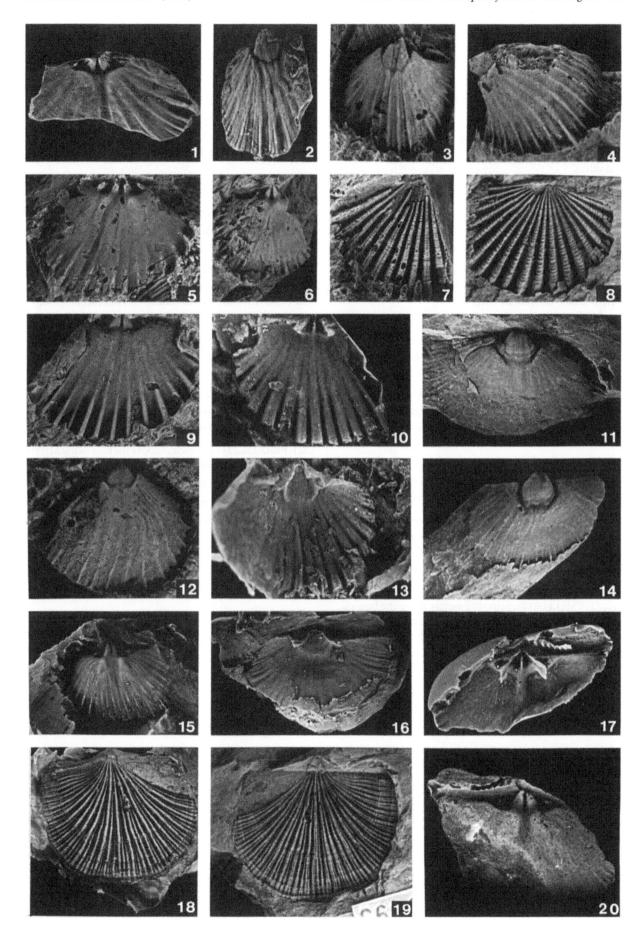

Plate 4

*1–2. ?Hesperorthis gwalia (Bancroft, 1949),
pp. 16–17.*

☐1. PMO 128.233. At the base of the Myren Member, Solvik Formation, Nesøya, Asker; ×3. Internal moulds of two brachial valves.

☐2. PMO 130.913. Counterpart of PMO 128.233; ×3. External moulds of two brachial valves.

3–11, 14. Eridorthis vidari sp. nov., pp. 17–18.

☐3. PMO 103.475. The upper parts of the Leangen Member, Solvik Formation, Skytterveien, Asker; ×3. Latex cast of internal mould of brachial valve.

☐4, 7. PMO 128.185. Sixty-five meters above the base of the Leangen Member, Solvik Formation Slemmestadveien, Leangbukta, Asker; ×3. Latex cast (4) and internal mould (7) of brachial valve.

☐5, 8. PMO 128.183. Sixty-five meters above the base of the Leangen Member, Solvik Formation, Slemmestadveien, Leangbukta, Asker; ×3. Latex cast (5) and internal mould (8) of brachial valve.

☐6. PMO 128.193. Sixty-five meters above the base of the Leangen Member, Solvik Formation, Slemmestadveien, Leangbukta, Asker; ×3. Internal mould of pedicle valve.

☐9, 14. PMO 128.188. Sixty-five meters above the base of the Leangen Member, Solvik Formation, Slemmestadveien, Leangbukta, Asker; ×3 (9), ×4 (14). External mould (9) and latex cast (14) pedicle valve. To the right notice latex of small brachial valve with narrow sulcus.

☐10, 11. PMO 128.188. Sixty-five meters above the base of the Leangen Member, Solvik Formation, Slemmestadveien, Leangbukta, Asker; ×5. Internal mould (10) and latex cast (11) of pedicle valve.

12. Glypthortinae sp., p. 18.

☐12. PMO 128.195. Thirty-six meters above the base of Myren Member, Solvik Formation, Spirodden, Asker; ×3. Internal mould of brachial valve.

13, 15–23. ?Plectorthis sp., p. 19.

☐13, 15. PMO 128.134. Nine meters above the base of the Myren Member, Solvik Formation, Konglungø, Asker; ×3. Internal mould (13) and latex cast (15) of brachial valve.

☐16, 17. PMO 128.100. Fifteen meters above the base of the Myren Member, Solvik Formation, Spirodden, Asker; ×3. Latex cast (16) and internal mould (17) of brachial valve.

☐18, 19. PMO 128.102a, b. Six meters above the base of the Myren Member, Drammensveien, Vakås, Asker; ×5. Internal mould (a) and counterpiece with external mould (b) of brachial valve.

☐20. PMO 128.149. Ten meters above the base of the Myren Member, Solvik Formation, Drammensveien, Vakås, Asker; ×5. Internal mould of pedicle valve.

☐21. PMO 128.151. Ten meters above the base of the Myren Member, Solvik Formation, Drammensveien, Vakås, Asker; ×5. Internal mould of pedicle valve.

☐22, 23. PMO 128.205. Nine meters above the base of the Myren Member, Solvik Formation, Drammensveien, Vakås, Asker; ×3. Latex cast (22) and internal mould (23) of pedicle valve.

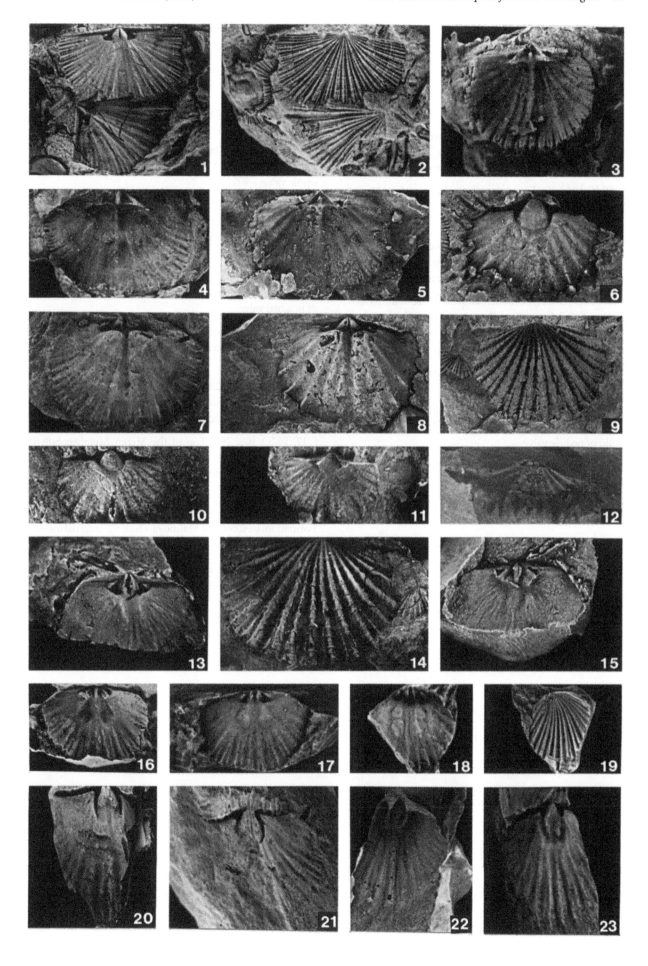

Plate 5

1–8, 11. Plectorthid indet., p. 19.

□1, 5. PMO 128.135. Nine meters above the base of the Myren Member, Solvik Formation, Konglungø, Asker; ×4. Internal mould (1) and latex cast (5) of brachial valve.

□2. PMO 128.136. Nine meters above the base of the Myren Member, Solvik Formation, Konglungø, Asker; ×3. Internal mould of brachial valve.

□3, 7. PMO 128.103. Fifty meters above the base of the Myren Member, Solvik Formation, Spirodden, Asker; ×3). Internal mould (3) and latex cast (7) of pedicle valve.

□4, 8. PMO 128.127. Six meters above the base of the Myren Member, Solvik Formation, Drammensveien, Vakås, Asker; ×4. Internal mould (4) and latex cast (7) of pedicle valve.

□6. PMO 128.130. Six meters above the base of the Myren Member, Solvik Formation, Drammensveien, Vakås, Asker; ×3. Internal mould of pedicle valve.

□11. PMO 139.137. Fifty meters above the base of the Myren Member, Solvik Formation, Spirodden, Asker; ×3. Internal mould of pedicle valve.

9, 10, 12–20. Platystrophia brachynota (Hall, 1843), p. 20.

□9, 13. PMO 139.138. Padda Member, Solvik Formation, Malmøya, Oslo; ×2. Ventral (9) and posterior (13) view of whole valve.

□10, 14. PMO 52.595. About 100 m above the base of the Myren Member, Solvik Formation, Spirodden, Asker, ×1.5 (10), ×2 (14). Ventral (10) and dorsal (14) view of whole valve.

□12. PMO 111.686. At the base of the Padda Member, Solvik Formation, west coast of Malmøya, Oslo; ×2. Internal mould of brachial valve.

□15. PMO 128.228. At the base of the Padda Member, Solvik Formation, west coast of Malmøya, Oslo; ×2. Internal mould of pedicle valve.

□16, 19. PMO 105.237. About 50 m above the base of the Leangen Member, Solvik Formation, Skytterveien, Asker; ×2. Latex cast (16) and internal mould (19) of pedicle valve.

□17. PMO 108. 283. Top of the Myren Member, Solvik Formation, west coast of Malmøya, Oslo; ×2. Internal mould of brachial valve. Refigured from Thomsen & Baarli (1982, Pl. 1:6).

□18. PMO 128.226. At the base of Padda Member, Solvik Formation, west coast of Malmøya, Oslo; ×2. Internal mould of brachial valve.

□20. PMO 128.230. The upper parts of Leangen Member, Solvik Formation, Kjørbo, Sandvika; ×2. Internal mould of brachial valve.

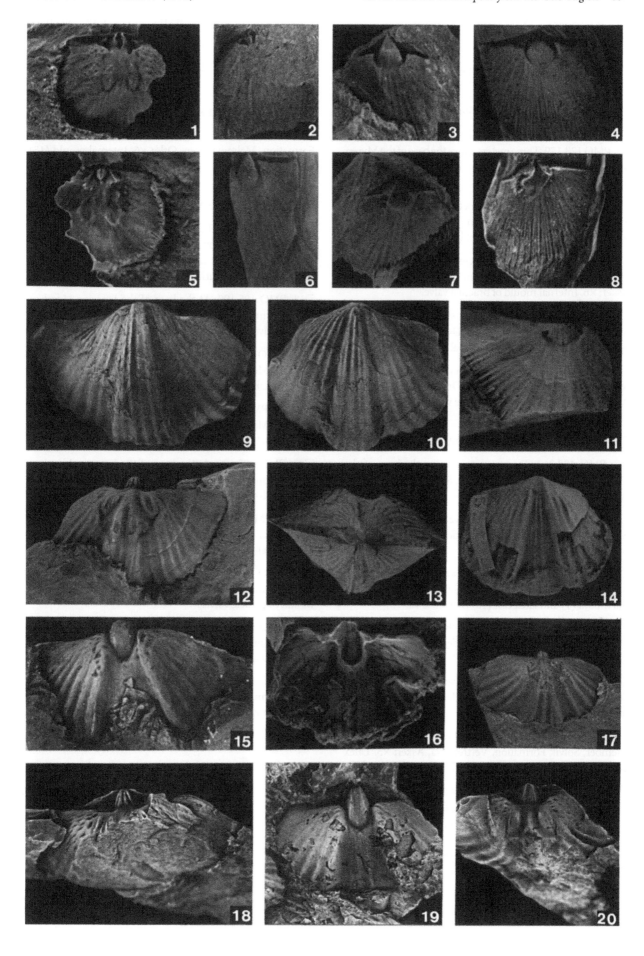

Plate 6

1–6 Skenidioides scoliodus *Temple, 1968,*
pp. 22–23.

□1. PMO 128.179. Three meters above the base of the Myren Member, Solvik Formation, Konglungø, Asker; ×5. Internal mould of pedicle valve.

□2, 6. PMO 128.222. Seventeen meters above the base of the Myren Member, Drammensveien, Vakås, Asker; ×4. Internal mould (2) and latex cast (6) of brachial valve.

□3. PMO 128.225. Seventeen meters above the base of the Myren Member, Solvik Formation, Drammensveien, Vakås, Asker; ×6. Internal mould of brachial valve.

□4. PMO 128.224. Seventeen meters above the base of the Myren Member, Solvik Formation, Drammensveien, Vakås, Asker; ×5. Internal mould of brachial valve.

□5. PMO 128.223. Seventeen meters above the base of the Myren Member, Solvik Formation, Drammensveien, Vakås, Asker; ×6. Internal mould of pedicle valve.

7–14. Skenidioides hymiri *sp. nov., pp. 23–24.*

□7. PMO 128.267. Twenty meters above the base of the Myren Member, Solvik Formation, Spirodden, Asker; ×8. Internal mould of brachial valve.

□8. PMO 128.170. Sixty meters above the base of the Myren Member, Solvik Formation, Spirodden, Asker; ×5. Internal mould of brachial valve.

□9, 13. PMO 128.169. Sixty meters above the base of the Myren Member, Solvik Formation, Spirodden, Asker. Internal mould (9) and latex cast (13) of pedicle valve.

□10. PMO 128.165. 104 m above the base of the Myren Member, Solvik Formation, Spirodden, Asker; ×5. Internal mould of brachial valve.

□11. PMO 128.177. Sixty meters above the base of the Myren Member, Solvik Formation, Spirodden, Asker; ×6. Internal and external mould of pedicle valve.

□12. PMO 103.490. 3.5 m above the base of the Leangen Member, Solvik Formation, Vettrebukta, Asker; ×5. Latex cast of the brachial valve figured by Thomsen & Baarli (1982, Pl. 1:4).

□14. PMO 103.491. Eight meters above the base of the Leangen Member, Solvik Formation, Vettrebukta, Asker; ×5. Internal mould of pedicle valve.

15, 16. Skenidioides *sp., p. 24.*

□15, 16. PMO 135.971. Ten meters above the base of the Vik Formation, Malmøykalven, Oslo; ×6. Internal mould (15) and latex cast (16) of brachial valve.

17–28. Skenidioides worsleyi *sp. nov., pp. 21–22.*

□17, 21. PMO 105.892. Lowermost part of Leangen Member, Solvik Formation, Skytterveien, Asker; ×5. Internal mould (17) and latex cast (21) of brachial valve. Refigured from Thomsen & Baarli (1982, Pl. 1:5).

□18. PMO 105.190. At the base of the Leangen Member, Solvik Formation, Skytterveien, Asker; ×5. Internal mould of pedicle valve. Refigured from Baarli (1987, Fig. 5j).

□19. PMO 105.877. Ten meters above the base of the Leangen Member, Solvik Formation, Skytterveien, Asker; ×5. Internal mould of brachial valve.

□20, 24. PMO 128.161. Ten meters above the base of the Leangen Member, Solvik Formation, Skytterveien, Asker; ×5. Internal mould (20) and latex cast (24) of brachial valve.

□22. PMO 105.191. At the base of the Leangen Member, Solvik Formation, Skytterveien, Asker; ×4. Internal mould of pedicle valve.

□23, 27. PMO 128.171. Sixty meters above the Myren Member, Solvik Formation, Spirodden, Asker; ×6. Internal mould (23) and latex cast (27) of brachial valve.

□25. PMO 136.004. At the base of the Myren Member, Solvik Formation, west coast of Malmøya, Oslo; ×4. Internal mould of pedicle valve.

□26. PMO 105.193a. At the base of the Leangen Member, Solvik Formation, Skytterveien, Asker; ×5. External mould of pedicle valve and counterpart of PMO 105.193b.

□28. PMO 105.193b. At the base of the Leangen Member, Solvik Formation, Skytterveien, Asker; ×5. Posterior view of internal mould of pedicle valve.

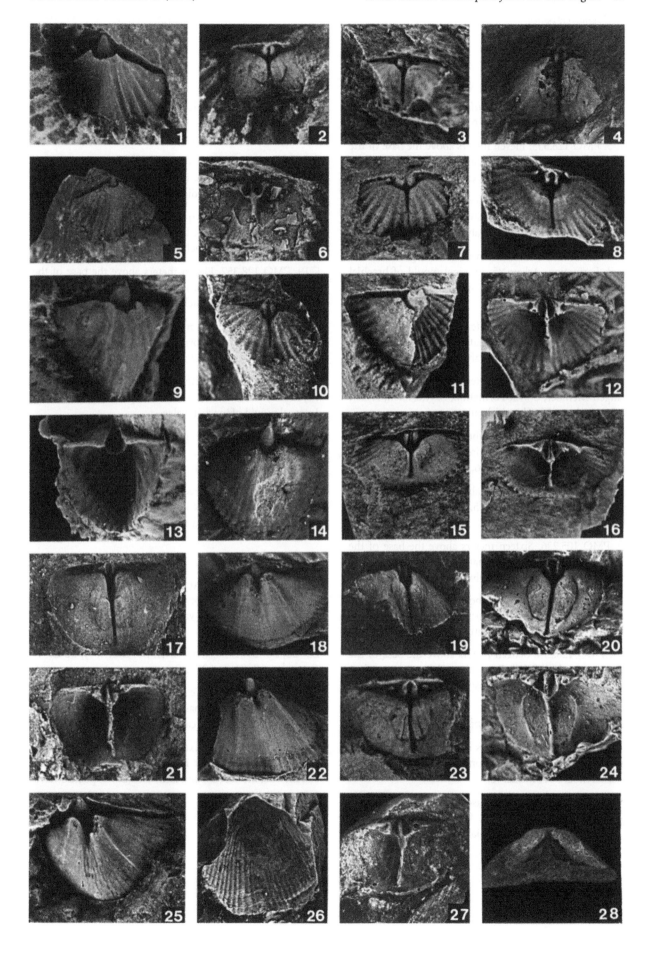

Plate 7

1–11. LEANGELLA SCISSA *(Davidson, 1871)*
TRIANGULARIS *(Holtedahl, 1916), pp. 24–26.*

☐1, 5. PMO 128.355. Eleven meters above the base of the Leangen Member, Solvik Formation, Skytterveien, Asker; ×5. Internal mould (1) and latex cast (5) of brachial valve.

☐2. PMO 128.344. Eleven meters above the base of the Leangen Member, Solvik Formation, Skytterveien, Asker; ×6. Internal mould of brachial valve.

☐3. PMO 105.225. Lower 20 m of the Leangen Member, Solvik Formation, Asker; ×5. External mould of brachial valve. From a slab with figured specimens in Thomsen & Baarli (1982, Pl 1:17).

☐4. PMO 105.226. Sixteen meters above the base of the Leangen Member, Solvik Formation, Skytterveien, Asker; ×5. Internal moulds of pedicle valves.

☐6. PMO 128.278. Sixteen meters above the base of the Leangen Member, Solvik Formation, Skytterveien, Asker; ×5. Internal mould of brachial valve.

☐7. PMO 107.601. Ten meters above the base of the Myren Member, Solvik Formation, Spirodden, Asker; ×5. Refigured from Baarli & Harper (1986, Pl. 2c).

☐8. PMO 128.266. Nine meters above the base of the Solvik Formation, Konglungø, Asker; ×5. Internal mould of pedicle valve.

☐9. PMO 105.226. Sixteen meters above the base of the Leangen Member, Solvik Formation, Skytterveien, Asker; ×5. Internal mould of pedicle valve. On same slab as Pl. 7:4.

☐10. PMO 128.371. Five meters below the base of the Rytteråker Formation, Leangen Member, Solvik Formation, Bleikerveien, Asker; ×5. Internal mould of brachial valve.

☐11. PMO 128.274. Sixteen meters above the base of the Leangen Member, Solvik Formation, Skytterveien, Asker; ×5. External mould of brachial valve.

12–23. AEGIRIA NORVEGICA *Öpik, 1933, pp. 26–27.*

☐12. PMO 128.340. Ten meters above the base of the Leangen Member, Solvik Formation, Skytterveien, Asker; ×6. External mould of brachial valve.

☐13, 17. PMO 128.362. At the base of the Leangen Member, Solvik Formation, Skytterveien, Asker; ×4. Internal mould (13) and latex cast (17) of brachial valve.

☐14, 18. PMO 128.356. Eleven meters above the base of the Leangen Member, Solvik Formation, Skytterveien, Asker; ×3. Internal mould (14) and latex cast (18) of pedicle valve.

☐15. PMO 128.363. Two meters above the base of the Leangen Memebr, Solvik Formation, Skytterveien, Asker; ×4. External mould of pedicle valve.

☐16. PMO 128.363. Two meters above the base of the Leangen Member, Solvik Formation, Skytterveien, Asker; ×4. Internal mould of brachial valve.

☐19. PMO 128.357. Eleven meters above the base of the Leangen Memebr, Solvik Formation, Skytterveien, Asker; ×4. External mould of brachial valve.

☐20. PMO 128.338. Ten meters above the base of the Leangen Member, Solvik Formation, Skytterveien, Asker; ×4. Internal mould of brachial valve.

☐21. PMO 128.347. Eighty meters above the base of the Myren Member, Solvik Formation, Spirodden, Asker; ×5. Internal mould of brachial valve.

☐22. PMO 105.195. Lower parts of Leangen Member, Solvik Formation, Skytterveien, Asker; ×4. Internal mould of pedicle valve. Refigured from Baarli (1987, Fig. 5e).

☐23. PMO 128.372. Three meters above the base of the Leangen Member, Solvik Formation, Skytterveien, Asker; ×4. Internal mould of pedicle valve.

24, 27. ?SOWERBYELLA *sp., p. 31.*

☐24, 27. PMO 105.210. At the base of the Myren Member, Ostøya, Asker; ×6 (24), ×9 (27). Internal mould (24) and latex cast (27) of brachial valve.

25–26, 28–29. EOPLECTODONTA DUPLICATA *(J. de C. Sowerby, 1839), pp. 27–29.*

☐25, 28. PMO 128.318. Near the base of the Padda Member, Solvik Formation, west coast of Malmøya, Oslo; ×3.5. Internal mould (25) and latex cast (28) of brachial valve.

☐26, 29. PMO 128.164. Sixteen meters above the base of the Leangen Member, Solvik Formation, Skytterveien, Asker; ×6. Internal mould (26) and latex cast (29) of pedicle valve.

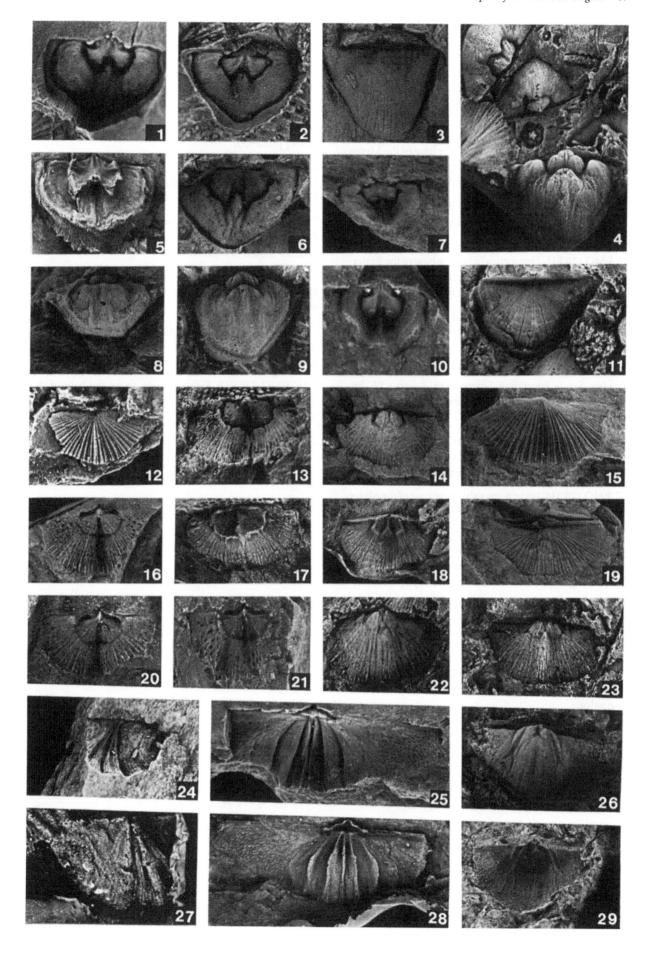

Plate 8

1–7, 10, 13, 16. EOPLECTODONTA DUPLICATA (J. de C. Sowerby, 1839), pp. 27–29.

☐1. PMO 128.300. Ten meters above the base of the Myren Member, Solvik Formation, Drammensveien, Vakås, Asker; ×4. Latex cast of external mould of valve.

☐2. PMO 105.881. Sixteen meters above the base of the Leangen Member, Skytterveien, Asker; ×4. Internal mould of brachial valve.

☐3. PMO 128.311. Five meters below the base of the Rytteråker Formation, in the Leangen Member, Solvik Formation, Skytterveien, Asker; ×5. Internal mould of brachial valve.

☐4. PMO 136.008. Near the base of the Myren Member, Solvik Formation, Spirodden, Asker; ×5. Latex cast of external mould.

☐5. PMO 128.268. Ten meters above the base of the Leangen Member, Solvik Formation, Skytterveien, Asker; ×3.5. External mould of brachial valve with abnormal thickening of growthlines.

☐6. PMO 128.320. Seventeen meters above the base of the Myren Member, Solvik Formation, Drammensveien, Vakås, Asker; ×4. Internal mould of pedicle valve.

☐7. PMO 128.314. Forty meters above the base of the Myren Member, Solvik Formation, Spirodden, Asker; ×6. Latex cast of external mould.

☐10. PMO 128.309. Five meters below the base of the Rytteråker Formation, in the Lengen Member, Solvik Formation, Skytterveien, Asker; ×4.5. Internal mould of pedicle valve.

☐13. PMO 128.315. Fifty meters above the base of the Leangen Member, Solvik Formation, Skytterveien, Asker; ×3. Pedicle valve.

☐16. PMO 128.288. Thirty meters above the base of the Myren Member, Solvik Formation, Spirodden, Asker; ×3.5. Internal mould of pedicle valve and external mould of brachial valve.

8–9, 11–12, 14–15, 17–18. EOPLECTODONTA TRANSVERSALIS (Wahlenberg, 1818) JONGENSIS subsp. nov., pp. 29–31.

☐8, 9. PMO 135.954. Uppermost parts of the Rytteråker Formation, Christian Skredsviks vei, Valler, Bærum; ×6. Latex cast (8) and internal mould (9) of brachial valve.

☐11. PMO 135.963. Uppermost parts of the Rytteråker Formation, Christian Skredsviks vei, Valler, Bærum; ×5. Internal mould of brachial valve.

☐12. PMO 130.936. Eighty meters above the base of the Rytteråker Formation, Kampebråten, Sandvika; ×4. Internal mould of pedicle valve.

☐14. PMO 111.665. Eighty meters above the base of the Rytteråker Formation, Kampebråten, Sandvika; ×6. Internal mould of pedicle valve.

☐15. PMO 130.935. Eighty meters above the base of the Rytteråker Formation, Kampebråten, Sandvika; ×7. External mould of brachial valve.

☐17, 18. PMO 135.912. Eighty meters above the base of the Rytteråker Formation, Kampebråten, Sandvika; ×6. Latex cast (17) and external mould (18) of brachial valve.

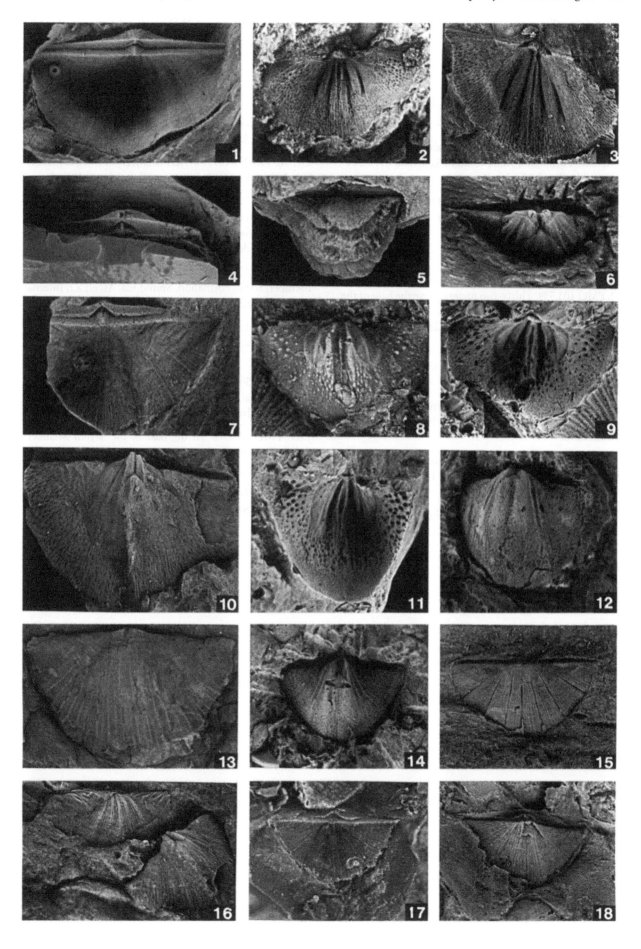

Plate 9

1, 4. EOPLECTODONTA TRANSVERSALIS (Wahlenberg, 1818) jongensis subsp. nov., pp. 29–31.

☐1. PMO 135.909. Seventy-six meters above the base of the Rytteråker Formation, Kampebråten, Sandvika; ×7. External mould of brachial valve.

☐4. PMO 135.958. Uppermost parts of the Rytteråker Formation, Christian Skredsviks vei, Valler, Bærum; ×6. Internal mould of brachial valve.

2. Plectambonitacea indet.

☐2. PMO 103.539. Thirty-two meters above the base of the Leangen Member, Solvik Formation, Slemmestadveien, Leangbukta, Asker; ×4.5. External mould of pedicle valve.

3, 5–18. KATASTROPHOMENA WOODLANDENSIS (Reed, 1917), pp. 31–32.

☐3. PMO 53.067. At the base of the Myren Member, Solvik Formation, Semsvannet, Asker; ×2. Internal mould of pedicle valve.

☐5, 8. PMO 130.355. At the base of the Myren Member, Solvik Formation, Nesøya, Asker; ×2. Internal mould (5) and latex cast (8) of pedicle valve.

☐6. PMO 130.357. Nine meters above the base of the Myren Member, Solvik Formation, Konglungø, Asker; ×2. Internal mould of brachial valve.

☐7, 10. PMO 130.360. Three meters above the base of the Myren Member, Solvik Formation, Konglungø, Asker; ×2. Internal mould (7) and latex cast (10) of brachial valve.

☐9, 16. PMO 40.097. Myren Member, Solvik Formation, Spirebukta, Asker; ×2. Posterior (9) and dorsal (16) view of valve. Refigured from Holtedahl (1916, Pl. 13:6).

☐11, 14. PMO 130.361. Three meters above the base of the Myren Member, Solvik Formation, Konglungø, Asker; ×3. Internal mould (11) and latex cast (14) of brachial valve.

☐12. PMO 130.353. Ten meters above the base of the Myren Member, Solvik Formation, Spirodden, Asker; ×2. Internal mould of brachial valve.

☐13. PMO 128.238. At the base of the Myren Member, Solvik Formation, Nesøya, Asker; ×2. Internal mould of pedicle valve.

☐15. PMO 50.068. Near the base of the Myren Member, Solvik Formation, Semsvannet, Asker; ×2. External mould of pedicle valve.

☐17. PMO 130.918. At the base of the Myren Member, Solvik Formation, Ostøya, Asker; ×3. Internal mould of pedicle valve.

☐18. PMO 51.976. At the base of the Myren Member, Solvik Formation, Semsvannet, Asker; ×2. External mould of brachial valve.

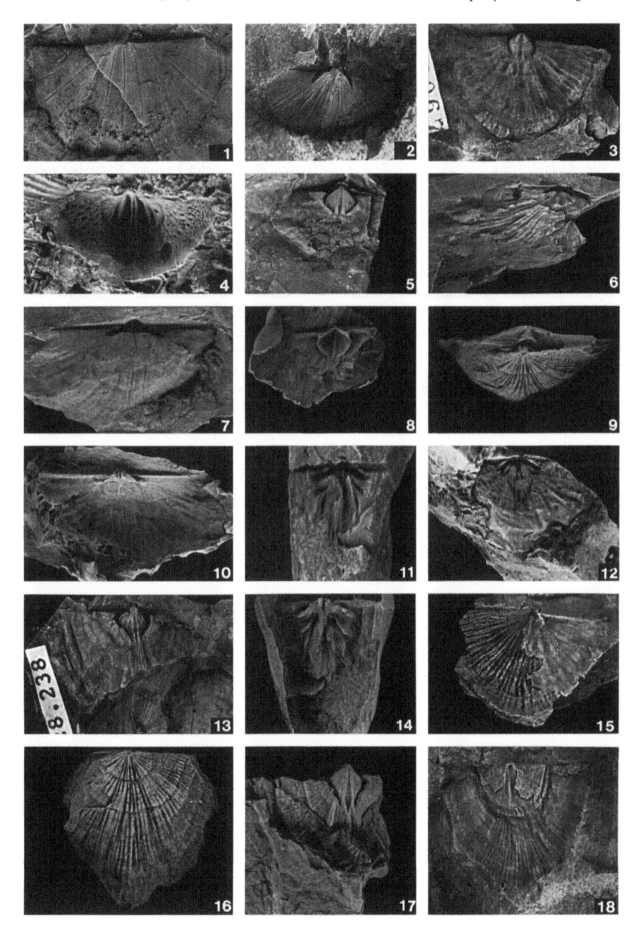

Plate 10

1–19. Katastrophomena penkillensis (Reed, 1917), pp. 32–34.

☐1. PMO 128.325. Eleven meters above the base of the Leangen Member, Solvik Formation, Skytterveien, Asker; ×3. External mould of valve.

☐2, 5. PMO 105.870a, b. Sixteen meters above the base of the Leangen Member, Solvik Formation, Skytterveien, Asker; ×2.5. Internal (2, a) and external (5, b) mould of pedicle valve from part and counterpart. The external mould is refigured from Thomsen & Baarli (1982, Pl. 1:12).

☐3, 6. PMO 128.323. Eleven meters above the base of the Leangen Member, Solvik Formation, Skytterveien, Asker; ×2. Internal mould (3) and latex cast (6) of brachial valve.

☐4. PMO 128.249. Ten meters above the base of the Leangen Member, Solvik Formation, Skytterveien, Asker; ×3. External mould of valve.

☐7. PMO 128.383. Two meters above the base of the Leangen Member, Solvik Formation, Skytterveien, Asker; ×4. Internal mould of pedicle valve.

☐8, 11. PMO 128.399. The uppermost parts of the Leangen Member, Solvik Formation, Skytterveien, Asker; ×2.5. Internal mould (8) and latex cast (11) of pedicle valve.

☐9, 12. PMO 105.885. Ten meters above the base of the Leangen Member, Solvik Formation, Skytterveien, Asker; ×2. Internal mould (9) and latex cast (12) of pedicle valve.

☐10, 13. PMO 128.240. At the base of the Leangen Member, Solvik Formation, Skytterveien, Asker; ×3. Latex cast (10) and internal mould (13) of brachial valve.

☐14, 17. PMO 128.324. Eleven meters above the base of the Leangen Member, Solvik Formation, Skytterveien, Asker; ×3. Internal mould (14) and latex cast (17) of brachial valve.

☐15, 16. PMO 130.911. Ten meters above the base of the Leangen Member, Solvik Formation, Skytterveien, Asker; ×3. Internal mould (15) and latex cast (16) of brachial valve.

☐18, 19. PMO 103.544. Eleven meters above the base of the Leangen Member, Solvik Formation, Skytterveien, Asker; ×2.5. Internal mould (18) and latex cast (19) of brachial valve.

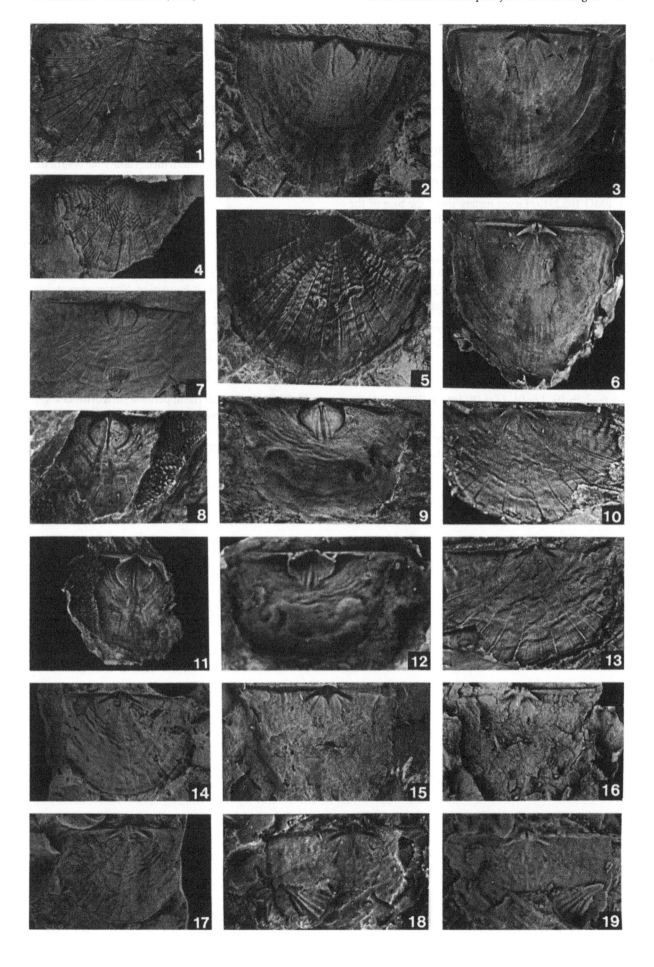

Plate 11

1, 5, 8, 11, 15–16. Dactylogonia dejecta sp. nov.,
pp. 34–35.

□1, 5. PMO 105.231a. Eleven meters above the base of the Leangen Member, Solvik Formation, Skytterveien, Asker; ×3. Internal mould (1) and latex cast (5) of brachial valve.

□8. PMO 105.231b. Eleven meters above the base of the Leangen Member, Solvik Formation, Skytterveien, Asker; ×2.5. External mould of brachial valve.

□11, 15. PMO 105.232. Eleven meters above the base of the Leangen Member, Solvik Formation, Skytterveien, Asker; ×2. Internal mould (11) and latex cast (15) of pedicle valve. Pl. 11:11 is refigured from Thomson & Baarli (1982, Pl. 2:3).

□16. PMO 128.365. Sixteen meters above the base of the Leangen Member, Solvik Formation, Skytterveien, Asker; ×2. Internal mould of pedicle valve.

2–4, 6–7, 9–10, 12–14, 17. Leptaena haverfordensis
Bancroft, 1949, pp. 36–37.

□2, 6. PMO 128.233. At the base of the Myren Member, Solvik Formation, Nesøya, Asker; ×2. Latex cast (2) and internal mould (6) of brachial valve.

□3. PMO 128.238. At the base of the Myren Member, Solvik Formation, Nesøya, Asker; ×2. Internal mould of brachial valve.

□4, 7. PMO 128.345. Nineteen meters above the base of the Leangen Member, Solvik Formation, Skytterveien, Asker; ×2. Internal mould (4) and latex cast (7) of pedicle valve.

□9. PMO 130.349. From 0 m to 5 m above the base of the Leangen Member, Solvik Formation, Vettrebukta, Asker; ×2. Pedicle valve.

□10. PMO 103.511. Seventy meters above the base of the Leangen Member, Solvik Formation, Skytterveien, Asker; ×3. Internal mould of pedicle valve. Refigured from Thomsen & Baarli (1982, Pl. 2:5).

□12. PMO 128.366. Sixteen meters above the base of the Leangen Member, Solvik Formation, Skytterveien, Asker; ×2.5. Internal mould of pedicle valve.

□13, 17. PMO 108.273. The uppermost parts of the Leangen Member, Solvik Formation, Skytterveien, Asker; ×3. Latex cast (13) and internal mould (17) of brachial valve. Pl. 11:17 is refigured from Thomsen & Baarli (1982, Pl. 2:2).

□14. PMO 128.234. At the base of the Myren Member, Solvik Formation, Nesøya, Asker; ×2. External mould of brachial valve.

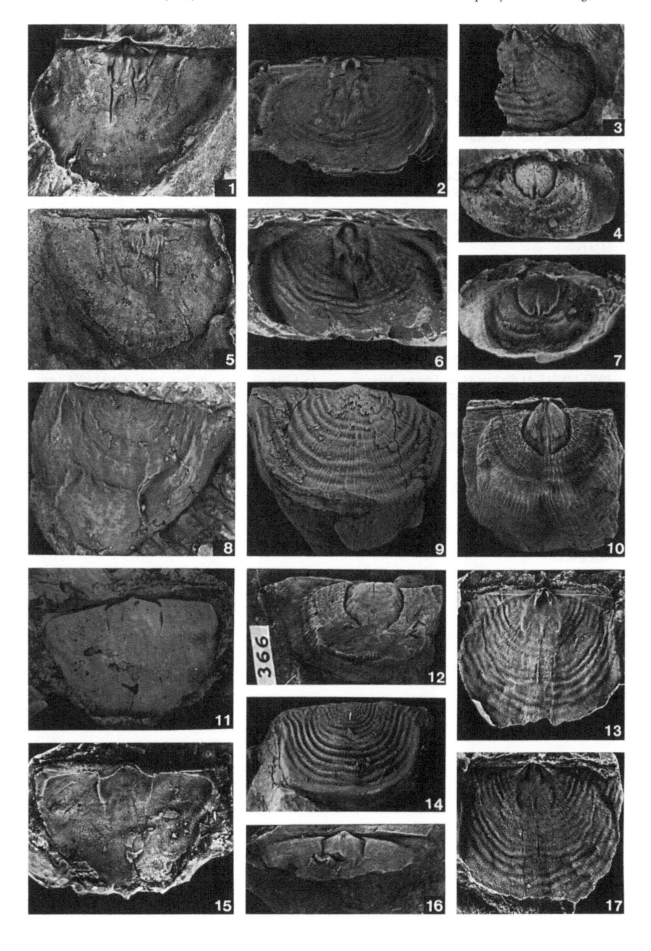

Plate 12

1–4. Leptaena valentia *Cocks, 1968, pp. 38–39.*

☐1. PMO 128.248. Padda Member, Solvik Formation, west coast of Malmøya, Oslo; ×1.5. Pedicle valve.

☐2, 3. PMO 128.381. Lower parts of Padda Member, Solvik Formation, west coast of Malmøya, Oslo; ×4. Latex cast (2) and internal mould (3) of brachial valve.

☐4. PMO 128.379. Lower parts of the Padda Member, Solvik Formation, west coast of Malmøya, Oslo; ×2.5. Internal mould of pedicle valve.

5–9, 11–12, 14. Leptaena valida *Bancroft, 1949, pp. 37–38.*

☐5, 6. PMO 130.348. From 0 to 5 m above the base of the Leangen Member, Solvik Formation, Vettrebukta, Asker; ×2.5. Internal mould (5) and latex cast (6) of brachial valve.

☐7. PMO 128.245. At the base of the Leangen Member, Solvik Formation, Skytterveien, Asker; ×2. External mould of pedicle valve.

☐8. PMO 128.257. Spirodden Member, Solvik Formation, Spirodden, Asker; ×3. Internal mould of pedicle valve.

☐9. PMO 128.350. From 0 to 5 m above the base of the Leangen Member, Solvik Formation, Vettrebukta, Asker; ×3. Internal mould of pedicle valve.

☐11. PMO128.251. From 0 to 5 m above the base of the Leangen Member, Solvik Formation, Vettrebukta, Asker; ×2. Pedicle valve.

☐12. PMO 130.915. From 0 to 5 m above the base of the Leangen Member, Solvik Formation, Skytterveien, Asker; ×3. Pedicle valve.

☐14. PMO 105.195. Eleven meters above the base of the Leangen Member, Solvik Formation, Skytterveien, Asker; ×2. Internal mould of pedicle valve. Refigured from Thomsen & Baarli (1982, Pl. 2:1).

10, 13. Leptaena *sp., p. 39.*

☐10, 13. PMO 128.246. Leangen Member, Solvik Formation, Bærum Sykehjem, Bærum; ×2. Internal mould (10) and latex cast (13) of brachial valve.

15. Leptaena purpurea *Cocks, 1968, p. 39.*

☐15. PMO 135.949. Eighty meters above the base of the Rytteråker Formation, Kampebråten, Sandvika; ×2. Internal mould of pedicle valve.

16, 17. Leptaenidae indet.

☐16, 17. PMO 111.696. Eighty meters above the base of the Rytteråker Formation, Kampebråten, Sandvika; ×2. Internal mould (16) and latex cast (17) of pedicle valve.

18. Crassitestella *gen. nov.* reedi *(Cocks 1968), pp. 39–40.*

☐18. PMO 128.237. At the base of the Myren Member, Solvik Formation, Nesøya, Asker; ×4. Internal mould of pedicle valve.

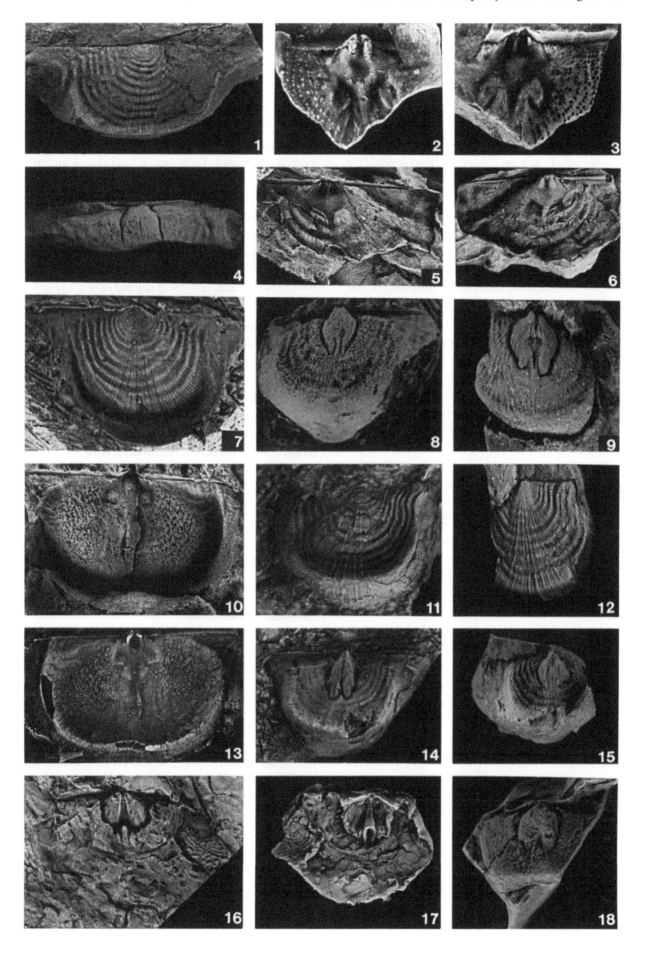

Plate 13

1–11. Crassitestella gen. nov. reedi (Cocks 1968), pp. 39–40.

☐1, 5. PMO 105.203a. At the base of the Myren Member, Solvik Formation, Brønnøya, Asker; ×4. Internal mould (1) and latex cast (5) of pedicle valve. Pl. 13:1 refigured from Thomsen & Baarli (1982, Pl. 2:8).

☐2, 3. PMO 130.919. At the base of the Myren Member, Solvik Formation, Brønnøya, Asker; ×4. Internal mould (2) and latex cast (3) of brachial valve.

☐4, 8. PMO 105.211. At the base of the Myren Member, Solvik Formation, Ostøya, Asker; ×3. Internal mould (4) and latex cast (8) of brachial valve. Pl. 13:4 refigured from Baarli & Harper (1986, Pl. 3f).

☐6, 7. PMO 105.203b. At the base of the Myren Member, Solvik Formation, Brønnøya, Asker; ×4. Internal mould (6) and latex cast (7) of brachial valve. On the same slab as Pl. 13:1.

☐9. PMO 130.920. At the base of the Myren Member, Solvik Formation, Brønnøya, Asker; ×4. External mould of pedicle valve.

☐10. PMO 128.260. At the base of the Myren Member, Solvik Formation, Ostøya, Asker; ×3. External mould of brachial valve.

☐11. PMO 130.921. At the base of the Myren Member, Solvik Formation, Ostøya, Asker; ×4. Internal mould of brachial valve.

12–23. Cyphomenoidea wisgoriensis (Lamont & Gilbert, 1945), pp. 40–41.

☐12. PMO 130.925. Eighty meters above the base of the Rytteråker Formation, Kampebråten, Sandvika; ×4. Internal mould of pedicle valve.

☐13. PMO 130.940. Eighty meters above the base of the Rytteråker Formation, Kampebråten, Sandvika; ×4. Internal mould of pedicle valve.

☐14, 15. PMO 130.942. Eighty meters above the base of the Rytteråker Formation, Kampebråten, Sandvika; ×4. Internal mould (14) and latex cast (15) of brachial valve.

☐16. PMO 130.928. Eighty meters above the base of the Rytteråker Formation, Kampebråten, Sandvika; ×3. Internal mould of pedicle valve.

☐17, 21. PMO 130.932. Eighty meters above the base of the Rytteråker Formation, Kampebråten, Sandvika; ×4. Latex cast (17) and internal mould (21) of brachial valve.

☐18, 19. PMO 108.272. Eighty meters above the base of the Rytteråker Formation, Kampebråten, Sandvika; ×4. Internal mould (18) and latex cast (19) of brachial valve.

☐20. PMO130.936. Eighty meters above the base of the Rytteråker Formation, Kampebråten, Sandvika; ×4. Internal mould of pedicle valve.

☐22. PMO 135.938. Eighty meters above the base of the Rytteråker Formation, Kampebråten, Sandvika; ×4. External mould of pedicle valve.

☐23. PMO 130.941. Eighty meters above the base of the Rytteråker Formation, Kampebråten, Sandvika; ×4. External mould of brachial valve.

24–26. Eostropheodonta multiradiata? (Bancroft, 1949), pp. 42–43.

☐24, 25. PMO 130.330a. 5–15 m below the base of the Rytteråker Formation in the Leangen Member, Solvik Formation, Skytterveien, Asker; ×4. Internal mould (24) and latex cast (25) of brachial valve.

☐26. PMO 103.467b. Uppermost parts of Leangen Member, Solvik Formation, Skytterveien, Asker; ×3. Internal mould of pedicle valve. On slab with material figured by Thomsen & Baarli (1982, Pl. 2:6).

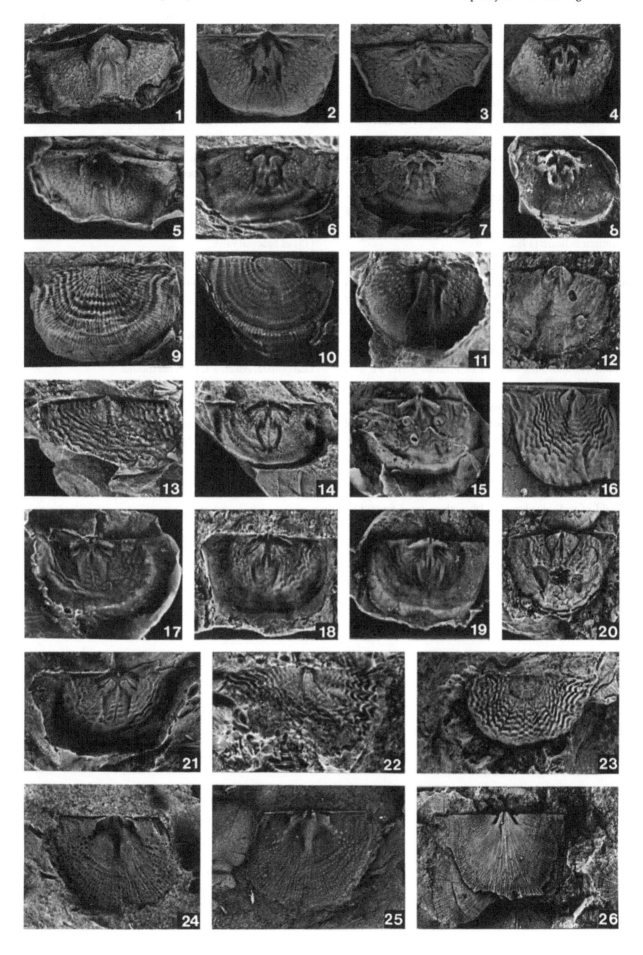

Plate 14

1–8. EOSTROPHEODONTA MULTIRADIATA*? (Bancroft, 1949), pp. 42–43.*

☐1. PMO 130.332. Upper half of the Leangen Member, Solvik Formation, Skytterveien, Asker; ×2. Exterior of pedicle valve.

☐2. PMO 103.467c. Uppermost parts of the Leangen Member, Solvik Formation, Skytterveien, Asker; ×2. Internal mould of pedicle valve. On slab with material figured by Thomsen & Baarli (1982, Pl. 2:6).

☐3. PMO 103.467d. Uppermost part of the Leangen Member, Solvik Formation, Skytterveien, Asker; ×4. Internal mould of brachial valve. On slab with material figured by Thomsen & Baarli (1982, Pl. 2:6).

☐4, 5. PMO 103.466. Uppermost part of the Leangen Member, Solvik Formation, Skytterveien, Asker; ×3. Internal mould (4) and latex cast (5) of brachial valve. Pl. 14:4 refigured from Thomsen & Baarli (1982, Pl. 2:4).

☐6. PMO 103.467e. Uppermost parts of the Leangen Member, Solvik Formation, Skytterveien, Asker; ×4. Internal mould of pedicle valve. On slab with material figured by Thomsen & Baarli (1982, Pl. 2:6).

☐7. PMO 103.467f. Uppermost parts of the Leangen Member, Solvik Formation, Skytterveien, Asker; ×2. Internal mould of pedicle valve. On slab with material figured by Thomsen & Baarli (1982, Pl. 2:6).

☐8. PMO 130.330b. From 5 to 15 m below the base of the Rytteråker Formation, in the Leangen Member, Solvik Formation, Skytterveien, Asker; ×4. Internal mould of brachial valve.

9–15. EOSTROPHEODONTA DELICATA *sp. nov., pp. 43–44.*

☐9. PMO 130.364. Forty-two meters above the base of the Leangen Member, Solvik Formation, Skytterveien, Asker; ×2.5. Internal mould of pedicle valve.

☐10, 13. PMO 128.376. Thirty-two meters above the base of the Leangen Member, Solvik Formation, Slemmestadveien, Leangbukta, Asker; ×3. Internal mould (10) and latex cast (13) of brachial valve.

☐11, 14. PMO103.470. Thirty-two meters above the base of the Leangen Member, Solvik Formation, Slemmestadveien, Leangbukta, Asker; ×4. Latex cast (11) and internal mould (14) of brachial valve.

☐12. PMO 128.377. Thirty-two meters above the base of the Leangen Member, Solvik Formation, Slemmestadveien, Leangbukta, Asker; ×4. External mould.

☐15. PMO 103.510. Thirty-two meters above the base of the Leangen Member, Solvik Formation, Slemmestadveien, Leangbukta, Asker; ×3. Internal mould of pedicle valve. On slab with *Dicoelosia osloensis* figured by Baarli (1988, Pl. 97:6).

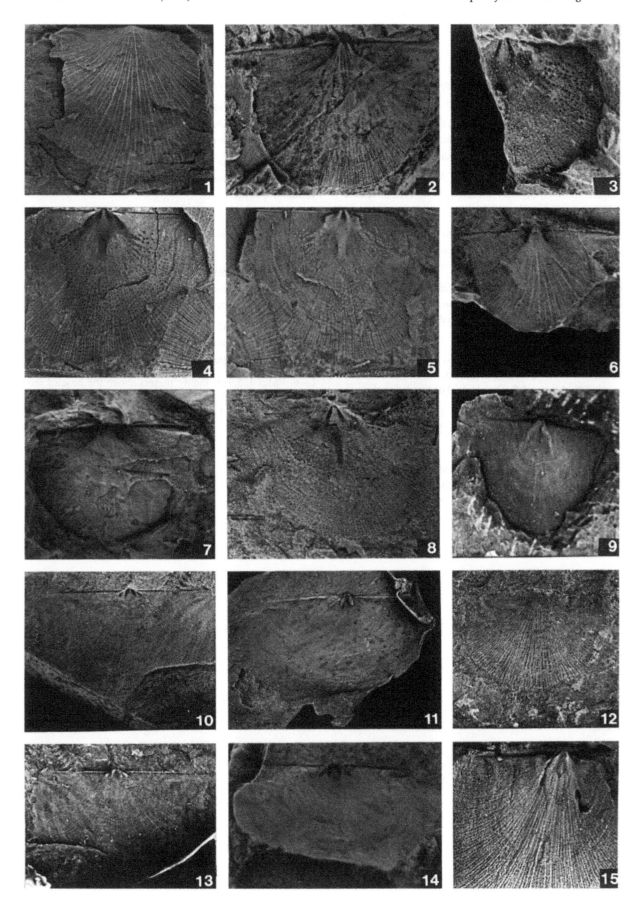

Plate 15

1–6, 8, 11, 14. Palaeoleptostrophia ostrina?
(Cocks, 1967), p. 44.

☐1. PMO 128.322. Thirty-two meters above the base of the Leangen Member, Solvik Formation, Slemmestadveien, Leangbukta, Asker; ×1.5. Internal mould of pedicle valve.

☐2. PMO 130.326. Sixty-two meters above the base of the Leangen Member, Solvik Formation, Skytterveien, Asker; ×4. Internal mould of brachial valve,

☐3. PMO 130.327. Six meters above the base of the Leangen Member, Solvik Formation, Skytterveien, Asker; ×4. Internal mould of brachial valve.

☐4, 8. PMO 130.344. Twenty meters above the base of the Leangen Member, Solvik Formation, Slemmestadveien, Leangbukta, Asker; ×2.5. Internal mould (4) and latex cast (8) of brachial valve.

☐5. PMO 130.340. Sixty meters above the base of the Leangen Member, Solvik Formation, Slemmestadveien, Leangbukta, Asker; ×1.5. The external of a pedicle valve.

☐6. PMO 130.324. Sixty-two meters above the base of the Leangen Member, Solvik Formation, Skytterveien, Asker; ×4. Internal mould of brachial valve.

☐11, 14. PMO 103.477. Forty meters above the base of the Leangen Member, Solvik Formation, Slemmestadveien, Leangbukta, Asker; ×1.5. Internal mould (11) and latex cast (14) of pedicle valve.

9, 12, 15. ?Mesoleptostrophia sp. indet., p. 45.

☐9, 12. PMO 135.943. Eighty meters above the base of the Rytteråker Formation, Kampebråten, Sandvika; ×3. Internal mould (9) and latex cast (12) of brachial valve.

☐15. PMO 135.938. Eighty meters above the base of the Rytteråker Formation, Kampebråten, Sandvika; ×4. Internal mould of pedicle valve.

7, 10, 13, 16. Mesopholidostrophia sifae sp. nov.,
pp. 46–47.

☐7, 10. PMO 130.938. Eighty meters above the base of the Rytteråker Formation, Kampebråten, Sandvika; ×3. Internal mould (7) and latex cast (10) of brachial valve.

☐13. PMO 135.987. Eighty meters above the base of the Rytteråker Formation, Kampebråten, Sandvika; ×3. Internal mould of juvenile pedicle valve.

☐16. PMO 135.913. Eighty meters above the base of the Rytteråker Formation, Kampebråten, Sandvika; ×2. External mould of pedicle valve.

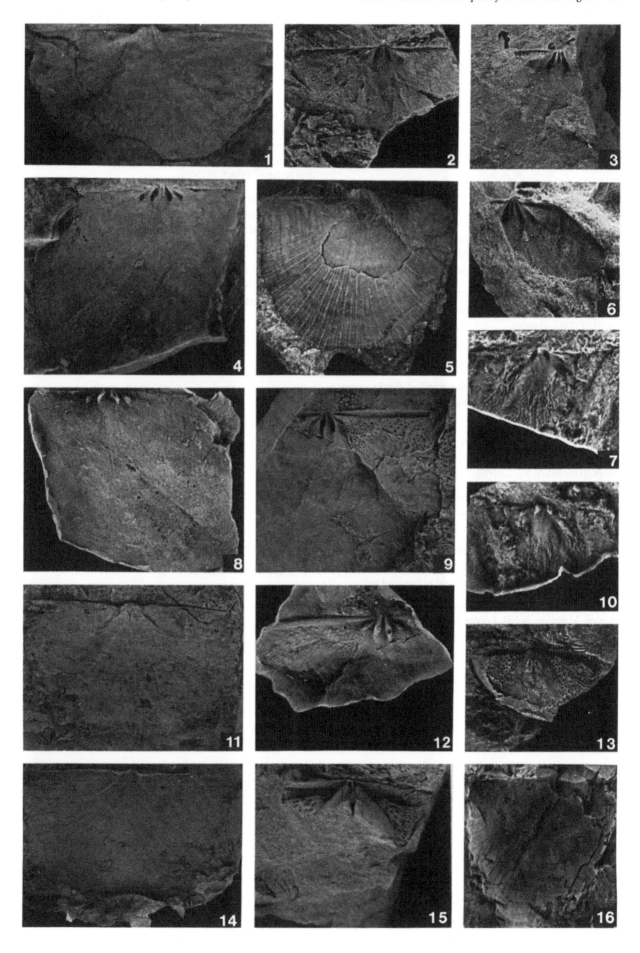

Plate 16

1–6. Mesopholidostrophia sifae sp. nov., pp. 46–47.

☐1, 4. PMO 108.266. Seventy-six meters above the base of the Rytteråker Formation, Kampebråten, Sandvika; ×5. Internal mould (1) and latex cast (4) of brachial valve.

☐2. PMO 135.932. Eighty meters above the base of the Rytteråker Formation, Kampebråten, Sandvika; ×3. Internal mould of pedicle valve.

☐3, 6. PMO 135.936. Eighty meters above the base of the Rytteråker Formation, Kampebråten, Sandvika; ×4. Internal mould (3) and latex cast (6) of brachial valve.

☐5. PMO 135.922. Eighty meters above the base of the Rytteråker Formation, Kampebråten, Sandvika; ×3. Internal mould of brachial valve.

7–12. Eomegastrophia spp.?, pp. 45–46.

☐7. PMO 128.391. The uppermost parts of the Leangen Member, Solvik Formation, Skytterveien, Asker; ×2. External of pedicle valve.

☐8. PMO 130.337a. Twenty-five meters above the base of the Leangen Member, Solvik Formation, Slemmestadveien, Leangbukta, Asker; ×2. Internal mould of pedicle valve.

☐9, 12. PMO 130.325. Sixty-two meters above the base of the Leangen Member, Solvik Formation, Skytterveien, Asker; ×2. Internal mould (9) and latex cast (12) of brachial valve.

☐10. PMO 128.400. Ten meters below the base of the Rytteråker Formation, in the Leangen Member, Solvik Formation, Skytterveien, Asker; ×2. Internal mould of brachial valve.

☐11. PMO 103.638. Sixty meters above the base of the Leangen Member, Solvik Formation, Skytterveien, Asker; ×2. Internal mould of pedicle valve.

13–15. Eopholidostrophia spp., p. 47.

☐13. PMO 105.229. Forty-two meters above the base of the Leangen Member, Solvik Formation, Skytterveien, Asker; ×4. Internal mould of pedicle valve. Refigured from Thomsen & Baarli (1982, Pl. 2:10).

☐14. PMO 105.228. Forty-two meters above the base of the Leangen Member, Solvik Formation, Skytterveien, Asker; ×4. External mould of pedicle valve. Counterpart of Pl. 16:13.

☐. 15. PMO 130.354. 80–85 m above the base of the Myren Member, Solvik Formation, Spirodden, Asker; ×3. Internal mould of pedicle valve.

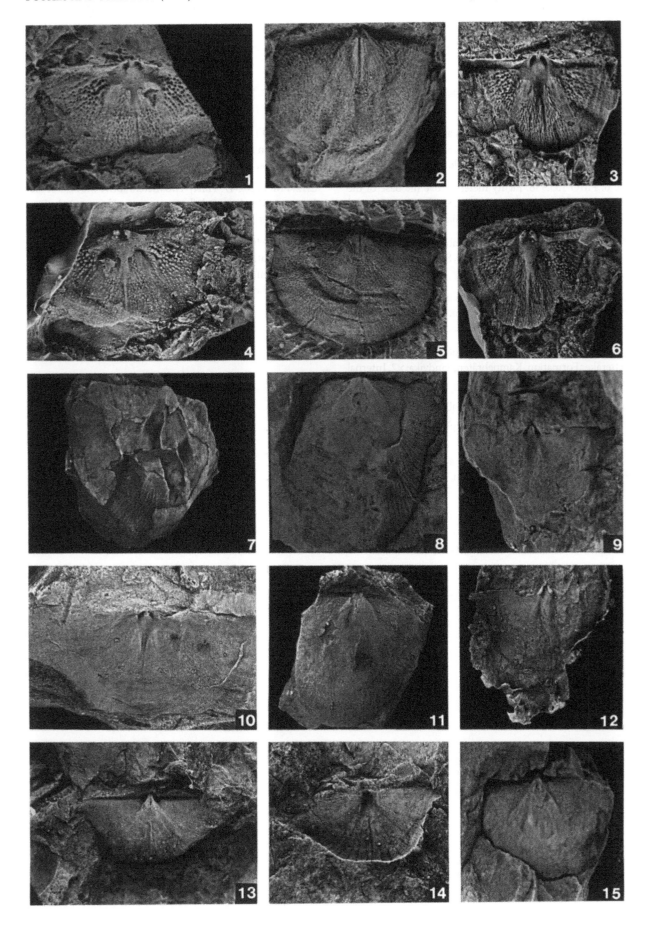

Plate 17

1–10, 13. EOCYMOSTROPHIA BALDERI gen. et sp. nov.,
pp. 48–49.

☐1, 2. PMO 135.945. Eighty meters above the base of the Rytteråker Formation, Kampebråten, Sandvika; ×2. Internal mould (1) and latex cast (2) of pedicle valve.

☐3, 6. PMO 135.956. Eighty meters above the base of the Rytteråker Formation, Kampebråten, Sandvika; ×3. Internal mould (3) and latex cast (6) of brachial valve.

☐4. PMO 135.944. Eighty meters above the base of the Rytteråker Formation, Kampebråten, Sandvika; ×1.5. Internal mould of brachial valve.

☐5, 7. PMO 135.968. Eighty meters above the base of the Rytteråker Formation, Kampebråten, Sandvika; ×2. Latex cast (5) and internal mould (7) of brachial valve.

☐8. PMO 135.970. Eighty meters above the base of the Rytteråker Formation, Kampebråten, Sandvika; ×2. Internal mould of pedicle valve.

☐9. PMO 135.942. Eighty meters above the base of the Rytteråker Formation, Kampebråten, Sandvika; ×2. Internal mould of pedicle valve.

☐10. PMO 135.935. Eighty meters above the base of the Rytteråker Formation, Kampebråten, Sandvika; ×2. External mould of brachial valve.

☐13. PMO 135.967. Eighty meters above the base of the Rytteråker Formation, Kampebråten, Sandvika; ×2. External mould of brachial valve. Counterpart to PMO 135.968.

11–12, 14–15. STROPHONELLA (EOSTROPHONELLA)
DAVIDSONI (Holtedahl, 1916), pp. 49–50.

☐11. PMO 130.337b. Twenty-five meters above the base of the Leangen Member, Solvik Formation, Slemmestadveien, Leangbukta, Asker; ×2. Internal mould of brachial valve.

☐12. PMO 130.347. Thirty-two meters above the base of the Leangen Member, Solvik Formation, Slemmestadveien, Leangbukta, Asker; ×1.5. Internal mould of pedicle valve.

☐14, 15. PMO 103.639. 5–10 m below the base of the Rytteråker Formation in the Leangen Member, Solvik Formation, Skytterveien, Asker; ×2. Internal mould (14) and latex cast (15) of pedicle valve.

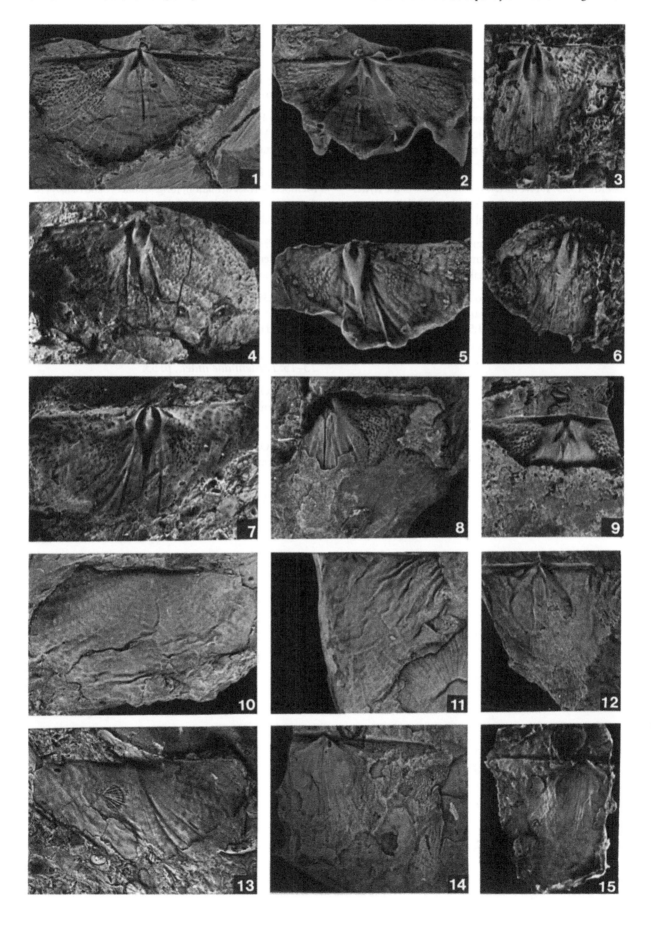

Plate 18

1–2, 4–5, 8, 11. Strophonella (Eostrophonella) davidsoni (Holtedahl, 1916), pp. 49–50.

□1, 4. PMO 128.367. 5–10 m below the base of the Rytteråker Formation, in the Leangen Member, Solvik Formation, Skytterveien, Asker; ×2. Internal mould (1) and latex cast (4) of brachial valve.

□2. PMO 130.371. Five meters below the base of the Rytteråker Formation, in the Leangen Member, Solvik Formation, Bleikerveien, Asker; ×2. External mould of brachial valve.

□5. PMO 130.345. Thirty-two meters above the base of the Leangen Member, Solvik Formation, Slemmestadveien, Leangbukta, Asker; ×2.5. Internal mould of brachial valve.

□8, 11. PMO 128.368. From 5 to 10 m below the base of the Rytteråker Formation, in the Leangen Member, Solvik Formation, Skytterveien, Asker; ×1.5. Internal mould (8) and latex cast (11) of brachial valve.

3, 6–7, 9–10, 12, 16. Saughina pertinax (Reed, 1917), pp. 50–51.

□3. PMO 103.472. Thirty-two meters above the base of the Leangen Member, Solvik Formation, Slemmestadveien, Leangbukta, Asker; ×3. Internal mould of pedicle valve.

□6, 7, 10. PMO 135.993. Uppermost part of the Leangen Member, Solvik Formation, Jongsaskollen, Sandvika; ×4 (6, 7), ×2 (10). Internal mould (6, 10) and latex cast (7) of brachial valve.

□9, 12. PMO 103.513. Ten meters above the base of the Myren Member, Solvik Formation, Spirodden, Asker; ×3. Internal mould (9) and latex cast (12) of brachial valve.

□16. 130.385. Three meters above the base of the Leangen Member, Solvik Formation, Skytterveien, Asker; ×2. External mould of brachial valve.

13–15. Fardeniidae indet., p. 53.

□13. PMO 130.367. Lower parts of the Padda Member, Solvik Formation, west coast of Malmøya, Oslo; ×3. Internal mould of pedicle valve.

□14. PMO 130.369. Lower parts of the Padda Member, Solvik Formation, west coast of Malmøya, Oslo; ×2. External mould of valve.

□15. PMO 130.344. Spirodden Member, Solvik Formation, Spirodden, Asker; ×3 External valve.

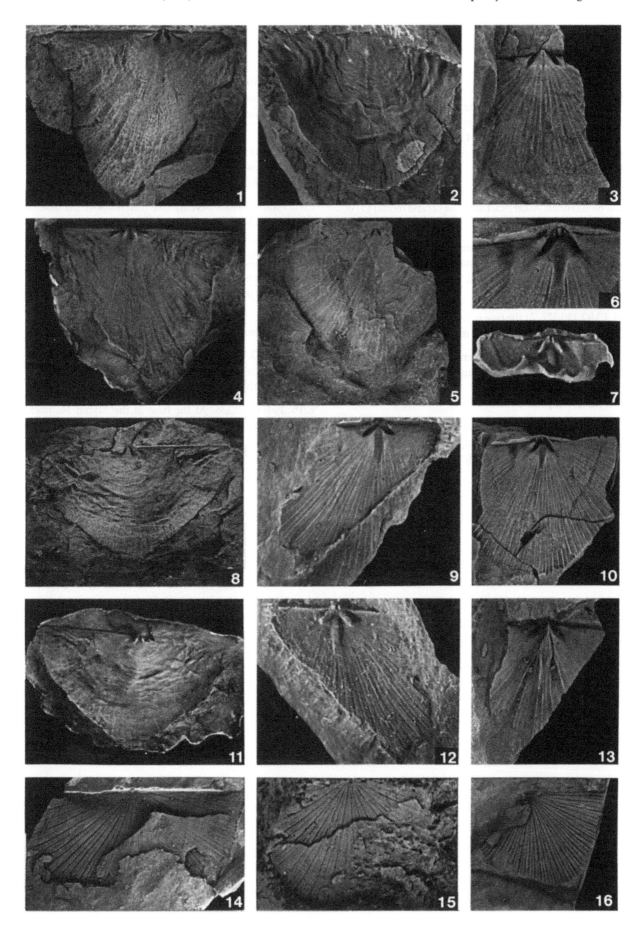

Plate 19

1–14. Fardenia oblectator sp. nov., pp. 51–53.

☐1. PMO 103.493a, b. Four meters below the base of the Rytteråker Formation, in the Leangen Member, Solvik Formation, Bleikerveien, Asker; ×2. Latex cast of brachial and pedicle valve. The moulds were figured by Thomsen & Baarli (1982, Pl. 2:9).

☐2, 3. PMO 130.370. At the base of the Rytteråker Formation, Bleikerveien, Asker; ×5. Latex cast (2) and mould (3) of exterior valve.

☐4. PMO 130.883. Seventy-five meters above the base of the Leangen Member, Solvik Formation, Slemmestadveien, Leangbukta, Asker; ×3. Internal moulds of pedicle and brachial valves.

☐5. PMO 130.377. At the base of the Rytteråker Formation, Bleikerveien, Asker; ×2. External mould of valve.

☐6, 8. PMO 130.890. Ten meters below the base of the Rytteråker Formation, in the Leangen Member, Solvik Formation, Skytterveien, Asker; ×3. Internal mould (6) and latex cast (8) of pedicle valve.

☐7. PMO 103.493c. Four meters below the base of the Rytteråker Formation, in the Leangen Member, Solvik Formation, Bleikerveien, Asker; ×3. Latex cast of mould of brachial valve.

☐9. PMO 103.512. Three meters below the base of the Rytteråker Formation, in the Leangen Member, Solvik Formation, Bleikerveien, Asker; ×3. Internal mould of brachial valve.

☐10, 11. PMO 130.370. At the base of the Rytteråker Formation, Bleikerveien, Asker; ×3. Internal mould (10) and latex cast (11) of brachial valve.

☐12. PMO 130.885. Seventy-five meters above the base of the Leangen Member, Solvik Formation, Slemmestadveien, Leangbukta, Asker; ×3.5. Internal mould of pedicle valve.

☐13, 14. PMO 109.725. Four meters above the base of the Leangen Member, Solvik Formation, Skytterveien, Asker; ×4. Internal mould (13) and latex cast (14) of pedicle valve.

Strophomenidae indet.

15–17. PMO 135.959. Near the base of the Vik Formation, Christian Skredsviks vei, Valler, Bærum; ×4. Internal mould of pedicle valve.

☐16. PMO 128.235. At the base of the Myren Member, Solvik Formation, Nesøya, Asker; ×4. Internal mould of pedicle valve.

☐17. PMO 130.356. At the base of the Myren Member, Solvik Formation, Nesøya, Asker; ×4. Internal mould of brachial valve.

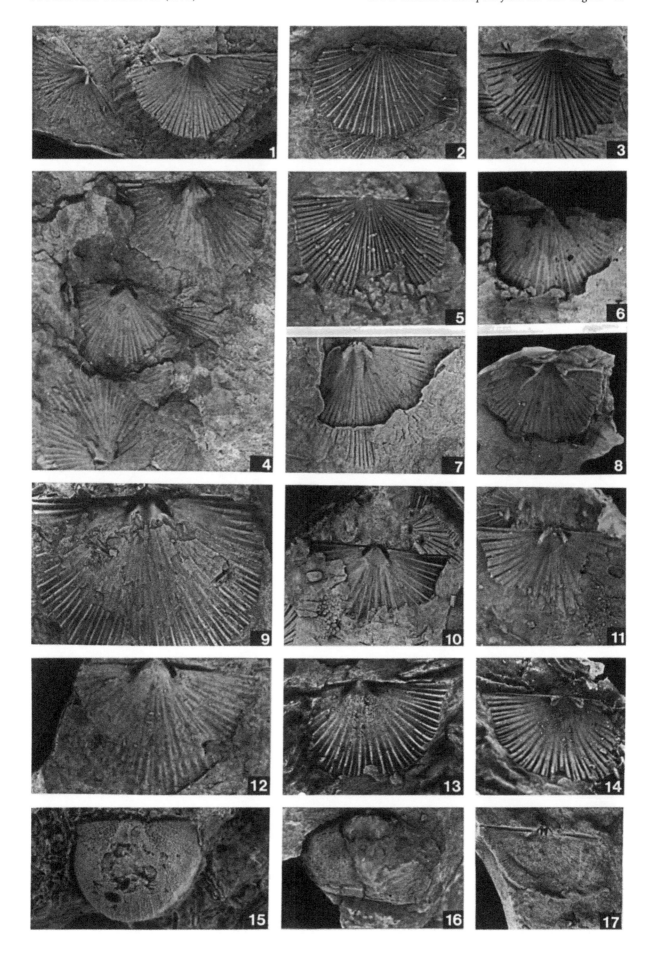

Printed and bound by CPI Group (UK) Ltd, Croydon, CR0 4YY

27/10/2024

14580393-0005